THE SCIENCE

OF THE

RACER'S BRAIN

THE SCIENCE OF THE RACER'S BRAIN

Otto Lappi PhD

Alan Dove

RB

Photo credits on p. 262

9 8

Categories
Science / Cognitive Science
Science / Life Sciences / Neuroscience

Sports & recreation / Motor Sports / Automobile Racing
Sports & recreation / Motor Sports / Motorcycle Racing

ISBNs

978-952-94-5878-3 (hardcover)
978-952-94-5854-7 (paperback)

Cover design by the authors

First published in 2022

www.racersbrain.org

DEDICATED

TO THE PURSUIT OF SPEED AND KNOWLEDGE

TABLE OF CONTENTS

PART III: THE BRAIN

PREFACE

ARE YOU CURIOUS about how your brain works? Since you have a brain (you are reading this, so you must do), you might be interested in finding out more about it. And it is, honestly, the most incredible piece of biological machinery you will ever come across. And if you have at some time watched an elite athlete perform, and you have asked yourself: *'How do they do that?'*, then you may also be interested in how some individuals are able to fine-tune this machinery, the same machinery that we all carry in our heads, to extraordinary levels. If you are interested in such things, then this book is for you.

You will find that it is a little bit different from other books about motor racing that you might have come across. In fact, it is not a 'book about motor racing' at all. Not fundamentally. It is a book about some core properties of the human mind, and about the way they get honed and developed to a very high degree in racing drivers.[1]

This book, or something very much like it, could have been written about many other sports as well: downhill skiing, freediving, mountain climbing… In any pursuit of excellence near 'the limit' – such an elusive goal! – many of the fundamentals, it appears, are the same. The preparation, the mental discipline, the 'feel' for one's environment, for one's equipment, and for one's body… and most of all the ability to maintain the clarity of mind that one needs to be able to evaluate a complex situation, and then make choices with life-affirming (or life-threatening) consequences, based on this evaluation. This is what the human brain does. This is how our brain is designed to operate 'on the edge'. And so, the underlying stories are universal.

But we two are first and foremost motorsport fans. For whatever reasons, we happen to love the particular sights, sounds and smells of this sport more than those of the others. So, this book was written to celebrate the ability to dance a 300 kph lump of plastic and metal with speed, power, precision – grace, even. It is the book we always wanted to read, but since nobody had written it, we had to do it ourselves.

WHY THIS BOOK IS NEEDED

When I (OL) was in London to meet Alan for this project, I went to the huuuuge four-story Waterstone's bookshop and was frankly amazed at the number of books on motorsport. There were so few!

And such as there were, were tucked away in a niche in the basement: a dedicated section probably smaller than my own modest racing book collection at home.

Elsewhere, historical leaders, poets, and celebrity chefs had their faces on covers. There were rows upon rows upon rows of books on money, health, sex, job satisfaction… and of course heaps of books all over the place that purported ('in a few simple steps!') to solve your problems with money, health, sex, job satisfaction…

There was also a great popular science section on the third floor, and it had many books on aviation, sports science, baseball… and of course a plethora of books on the neuroscience of this, that and the other thing. But no 'Neuroscience of Racing'.

We feel that a scientifically informed book about expert-level race driving and its cognitive and neural basis, is viable, timely and interesting. We also decided pretty early on that we want a book that is scientifically respectable; based on more than a superficial insight into cognition and brain function. This is why it is part-written by a scientist actively engaged in research – somebody who has dedicated their professional life to the pursuit of knowledge.

But at the same time we felt that it should be about *real racing* – as seen and felt from the inside. So, it also needs to be part-written by someone who has dedicated their life to the pursuit of speed.

And then we wanted the end product to be seamless, for both views to be combined into a whole. Luckily, we found that despite our very different background, education and personality, we shared the same 'language' – similar ideas, similar interests and similar ideals. In the end, there are many parts where we cannot even tell ourselves who wrote the first draft. This is how it should be.

What This Book Is…

Over the next two hundred pages we are going to break down the skills of the racing driver, analyze the process of driving, and dissect properties of the human mind and brain that make it possible. We intend to give you an accurate insight into some of the fascinating and complex brain mechanisms of a skilled racing driver.

We will take you on a tour of the mind and the brain, one that will give you a completely new of view of motor racing – through the lens of the underlying brain functions. We will combine the science of the mind and the brain with nitty-gritty details of driving technique – and blend them with the passion and sheer excitement of motorsport. This is information that you will not get from anywhere else.

...AND WHAT IT IS NOT

This is not a 'go-faster' book that will teach you how to drive a racing car. We want you to understand some of the principles and techniques, and how the racing driver's mind operates – and what happens in the brain when you are going really fast. But let's be real here: understanding these things, at a theoretical level, is not going to make you fast. (Not 'In a few simple steps!', certainly.) If you are a racing driver and want to go faster, this is not something you pick up from a bunch of tricks in a book, or by 'unlocking' some obscure hidden power of the brain.

For sure, thinking about your own performance, and reflecting on what you are doing right and what you are doing wrong, is really important. It's just that the understanding you gain from books and reading will not in and of itself lead to performance gains. You need track time, and preferably expert instruction to speed up the learning process. (**Chapter 5** and **Chapter 9** will explain why your brain design makes it so.)

This also calls for a disclaimer. We describe here many techniques of how to operate a racing car, or bike. This is done to illustrate scientific ideas – it is not offered as advice or suggestions. Any interpretations and applications you might make are your own responsibility. We cannot accept liability for your actions – or for your accidents.

WHO IS THIS FOR?

We write for everyone who appreciates the thrill of speed, and the skill involved in doing it well. We will not assume any prior knowledge of race driving techniques, or of neuroscience. Anyone who wants to understand how the brain works, and how this translates to real-world expert performance, should be able to follow all of the ideas. Technical jargon will be avoided wherever possible. We have made every effort to make this journey of discovery engaging, exciting and fun.

Within that, we will not be pulling punches in delving into some pretty complex subject matter. So, we will occasionally ask for some measure of effort and concentration on your part. But then, the topic *is* the human brain, the most complex object on the planet, operating sophisticated machinery at the highest levels of skill... So, realistically, what can you expect?

And we do want to be able to provide enough detail, nuance and insights to engage and inform motorsport professionals and the hardcore race fans, who want to understand all aspects of this sport.

(You will find extensive references to scientific literature at the end notes, which also give added technical detail – none of this is essential to enjoy and understand the rest of the book, and can be skipped. More free content is available on the book's website www.racersbrain.org.)

If you are a racing professional, such as a driver, engineer, coach, or a personal trainer, then the information contained here may not necessarily change the way you go about your daily business...but it might just change the way you think about things.

If you are a race fan, we hope that by the time you finish this book you will have a new appreciation of the human element between the seat and the steering wheel. We want to make you see things that you thought you knew in a completely new way.

And if you've picked up the book even though you're not that interested in racing as such, then *that's great!* We think you also will gain a lot.

Many of the principles are general; and we believe that by looking at a concrete example, concentrating on a single domain, will give you a clearer idea about how the brain works than what a general 'science of human performance' approach might. Because concrete examples *that all connect to one another* must fit together and interlock into a single coherent whole in a way that is less fluffy, less vague, and more 'complete' than a mere list of general principles or a hodge-podge of anecdotes from a menagerie of unrelated skills would be.

This way of looking at things is also useful if you are a student in behavioral, biological or sports sciences, and you want to understand human performance and its psychological and neuroscientific basis. Here you will do well to develop insights into how the brain works in concrete domains. And doing this first for one domain, such as racing, you should subsequently find it easier to work out for yourself how the brain might work in other domains that are of interest to you. May you have the freedom to pursue that interest.

If you are already a researcher in psychology or neuroscience or some other related field, and you are in principle intrigued about expert performance 'in the wild', you might justifiably feel that the scientific understanding of the mind and brain is not quite yet mature enough to describe such complex real-world skills. Things people actually do, outside the laboratory, and the way they do it, is just so complicated. So much is yet unknown; not many things can be known *for sure*. Well, we are there with you. Part of the reason for writing this book is to inspire research, and to move forward the discussion regarding driving (on both on the academic and on the racing side of things).

Maybe we are being a bit egotistic to embark on such a mission; but, according to Hemingway, exactly this accounts for many books! And surely, we thought, we could make a worthwhile contribution through bringing some of this stuff together, for the first time. Why not? What's the worst that can happen? How well we have been able to pull this off, on that you must be the judge.

Come the end, we can't guarantee to have all the answers ready. And of course, we realize that not everything we say is necessarily so. We try our hardest to describe the techniques of racing and the science behind them in an impartial, informed way. For any errors and biases that remain, we must accept blame. And even on things that are well established there can always be mistakes (if you spot any, we'd love to hear from you). And for sure, there are bound to be places where we are just trying our best to say something intelligent about things that are still deeply, profoundly unknown…

This book is just one small step into that unknown. Thank you for reading it.

‘A RACING DRIVER IS A RACING DRIVER. SOME ARE FASTER THAN OTHERS, AND THE INGREDIENT THAT MAKES ONE PERSON FASTER THAN ANOTHER (...) IS A MENTAL QUALITY WHICH, AS FAR AS I KNOW, HAS NO NAME.’

- SIR STIRLING MOSS [2]

PART I

THE SKILL

Chapter One

Observing the Racing Driver

THE BLARE OF A STRAIGHT-EIGHT RACING ENGINE ricochets from buildings, as a silver sports racing car rushes towards the finish line. There are two exhausted men in the cockpit. One of them, a little bearded man, waves to the crowd. Both he and the driver are wearing white helmets; the driver has racing goggles; the navigator just has eyeglasses. The exposed areas of their faces are covered in a thick, brown grime of sweat, dust, oil, rubber, pollen and flies. Deafened by the noise from the side exhaust of their Mercedes-Benz racecar, fatigued from the heat, and strained by the tension from the risk to their lives they have had to take that day, they've reached the finish.

The date is May 1st, 1955. The place is Brescia, in Italy. Big red letters on the bonnet indicate the car's starting number: 722. This also gives the car's starting time: 7.22 AM. Cars had set off at intervals of a few minutes, their starting number matching the start time so that spectators could know who's gaining ground and who's losing it. It is now 5.29 PM. In 10 hours and 7 minutes, Stirling Moss and his navigator Denis Jenkinson have covered 992 miles (1597 km) across Italy. And, although they cannot know it just yet, as cars that had set off after them are still to arrive, in doing so Moss and 'Jenks' have set a record that will stand forever. For over ten hours they have averaged – yes, averaged – 98 mph on the highways, back roads and mountain passes of northern Italy. The motor race was called the Mille Miglia.

Half a century after the supposed end of great city-to-city races, with motor racing now largely confined to permanent, purpose-built circuits, this was still a race like no other. It was run on public roads closed off for the event (well, in theory[3]); a 1600 km circuit from Brescia down to the Italian Riviera, to Rome, and then up through Tuscany back to Brescia. Can you imagine it? Hanging your laundry out to dry, or having your morning coffee in a sleepy Italian village – and just meters away, passing by in a blur of noise and colour and speed are the top Grand Prix drivers of the day, including world champions past, present and future, piloting some of the most powerful and sophisticated racing cars ever. Full factory efforts from the likes of Mercedes, Ferrari and Maserati going all-out. Maybe as many as five million spectators lined the roads that day.

For people of the time, a time of post-war austerity, it must have been like watching something from another dimension. And in a way it was; because the two men in the car would indeed enter a kind of separate mode of existence, travelling at double, triple, quadruple the normal traffic speeds…

The Mercedes-Benz 300 SLR endurance racing car, which was basically a toughened-up two-seater version of the 1955 Mercedes W196 Grand Prix car, was geared to over 170 mph, at which speed Moss and Jenkinson were 'cruising' at the early stages of the race, on their way to Verona, when they were caught up and passed by Eugenio Castellotti's Ferrari! (Castellotti crashed out of the race a little later.)

Once they'd come far enough at the front, or near enough to the front to be able to push for victory, Moss would have begun 'dicing'. 'Dicing with death', as Jenkinson called it.[4] This should be taken literally. A race such as this was not only a test of endurance for the machines, but of the willpower, mental stamina and resolve of the drivers as well.

Even fifty years later, Moss would admit that the Mille Miglia race actually frightened him. Because '*I knew how fast I had to go, and how little I knew about where I was going… because, you know, there was no way I was going to learn the circuit!*'[5] This is because a thousand miles across Italy is way too much to commit to memory – even for a super-elite[6] driver like Moss.

For this reason, Jenkinson had earlier devised with John Fitch[7] a navigational aid system called, internally, irreverently, 'the toilet roll'. This was a small, gray rectangular metal box that you could hold in your hands, about the size of an alarm clock. It had an aperture on one side, through which you could see a small part of the roll of paper inside. The paper could be moved up and down with two turn knobs on the right-hand side. On it, there were written instructions for intersections, how to take corners, crests, villages, or other situations requiring judgement. The recorded information would be terse and to the point, meant to be read out one line at a time, such as: 'long right' or 'blind left' or 'flat out between houses'.

There was no intercom like modern rally cars have, and the pilot and navigator could not hear one another over the roar of the engine and wind noise. Moss and Jenkinson had a system of hand signals to communicate the navigational instructions (as well as other important information, such as observations of competitors ahead or behind).

The pace notes had been prepared during reconnaissance. The Mercedes-Benz entry was a proper effort, dedicating three *months* to on-road testing and development in Italy, with Mercedes-Benz 220s, 300SLs and two 300 SLR racecars for practice. Moss and Jenkinson had covered 10 000 miles, going over and over the course from Brescia to Rome and back. This was not at racing speeds, but in a touring car, and *'for lots of the time we never went over 85 mph'*. This was before the time of speed limits, obviously...[8]

Over these more than one hundred hours, Jenkinson had the opportunity to observe first-hand the way Moss – one of the greatest racing drivers of all time[9] – would touch the controls. The speed and accuracy of his movements, the fluidity and balance of his control of the car; and also the way the car would then respond to him in its movements down the road. And then there were the more intangible powers of observation and situational awareness; the way Moss would 'read' the road scene and assess the information.[10]

In 1959, Jenkinson went on to write a little book about the psychological and physiological make-up of an elite racing driver.[11] It was based on his observations riding with Moss, and in sidecar motorcycle competition. It contained some detailed engineering insights into vehicle dynamics, and a contemporary understanding of sensory physiology, brain function, and has what since become known as cognitive psychology.

60 years on, it is still one of the most – perhaps *the* most – thoughtful, technically detailed and insightful pieces of writing on the topic. That is: the topic of identifying and analyzing the skills of individuals who are able to control a high-speed land vehicle at the highest level of ability. The unnamed 'mental quality' that makes one racing driver faster than another. It is about this mental quality that we also want to tell you about.

We are here to ask: can we understand now, from a modern cognitive science perspective, and with the benefit of 60+ years into the psychology and neuroscience of human performance, something new? Can we gain further insight into some of the things? We think we can – and can only hope that the information we provide would have interested and delighted the little bearded man himself.

This book is intended to give, as far as possible in the current state of knowledge, a look into the complex mechanisms in the mind and brain of a skilled racing driver. In short, we will try to give reasonable answers to the following:

- What skills does an elite racing driver possess?
- What cognitive functions underlie those skills?
- What brain mechanisms support such cognitive functions?

These are the topics of the three parts of the book. **Part I** is about getting clear on the job description for the racer's brain – what the skill involved actually consists in. **Part II** is about the basic ingredients of the mind involved – the cognitive basis of expert performance. **Part III** is about how fundamental mechanisms in the brain get the job done.

WHAT WE TALK ABOUT WHEN WE TALK ABOUT DRIVING

We will soon be making some observations on, and analyzing the techniques involved in operating a racing car. For those readers who are not into racing, much of this will be new, and for them we want to clear the scene a little bit. We do this by pointing out ways that expert performance in driving relates to your everyday performance – or doesn't.

Expert performance, in any field, often has a similar air of effortlessness as everyday routine has. And still, when you see athletes in many sports do their thing – even if it's on TV – then, you know, really know in your gut, that you absolutely, positively could not do what they do. Even as a casual observer, your brain recognizes that the way they are able to move their body is beyond your ability. You could not move your legs like an Olympic sprinter. You could not scale a sheer rock face like a freeclimber. You could not dive a hundred feet on a single draw of breath.

But then you hop channels and there's a motor race on. And, for many people, the brain will go: *Well, you know, I could probably do that – they're just sitting there and turning the wheel, what's the big deal?* When the casual observer looks at a racing car at speed, the natural thing is to relate it to your everyday experience: *More of the same, just at a higher level of speed (and risk),* the brain says. The tiny cues that tell how the vehicle is being balanced, how it is manipulated to extract maximum speed, may be too subtle to register to the non-expert eye.

And here it is not very helpful that in modern car racing the driver is strapped so tightly to the chassis that she[12] can only make minimal movements of the hands and the feet. And while modern racing cars are spectacular in their engineering, they are perhaps less spectacular to watch (especially on TV). Advances in tyre, suspension, and aerodynamic technology mean that the speeds are higher, but cars can appear to be 'cornering on rails'. All this further denies the casual viewer's brain the clear visual information that something unusual is happening.[13]

Thus, there is a problem in getting some of the information across. Not in spite of the reader's driving experience but *because* of it. On the one hand, it makes the domain familiar and relatable… but it can also lead to missing important differences. Some of these are presented in **Table 1.1.** (overleaf).

To get a feel for what is involved – and to guide the reader towards the right sorts of observations and intuitions – we will next describe one racing driver preparing and executing a single qualifying lap.

A few caveats. First, this is a fictional account, meant to introduce in a concrete and relatable way many basic concepts and techniques of race driving to the non-racer readers – and at the same time some basic psychological and neuroscience ideas to the racer readers. So, it leaves out a lot of 'real-world stuff', such as money, politics etc. It is meant to describe the 'pure essence' of the job at hand.

Second, we won't begin to try to convey the violence of driving, meaning the amount of brutal shaking and vibration and the physical sensations and emotions you get from putting yourself in harm's way. Some while ago, I (AD) took a 250 National kart out for a spin. This thing is bonkers; the vibration alone from the CR250 engine as you sit on the grid is downright *obscene!* The track was a little damp, and I am here to tell you, it can get sketchy as f*ck out there… Okay, you get the point. So, with that off our chests, picture this…

TABLE 1.1. EVERYDAY DRIVING VS. EXPERT RACE DRIVING

Everyday driving	Race driving
Steering wheel turned slowly and to large angles	Steering wheel turned quickly, and range is quite limited
Brake pedal is soft and pressed only very lightly	Brake pedal is very stiff and pressed with a lot of force
Throttle pedal is used vaguely, throttle control makes little difference	Accurate throttle control is critical for traction and balance
Little noise and vibration, negligible g-forces; comfortable temperature	A lot of noise, vibration & shocks, strong g-forces; heat stress
The traffic situation, other road users and potential hazards are all monitored	Focus is entirely on performance
Route and scene geometry unfamiliar, or known only very roughly	Track and curve geometry known in great detail
Time and opportunity to multitask (e.g. listening to music, watching the scenery, socializing…)	Focus on the driving task (might involve radio communications and managing car settings, though)
Speed, kinetic energy and forces well within the range the tyres can handle	Speed, kinetic energy and forces at the limit of the design of the vehicle
Enjoyable drives tend to be relaxing and pleasant ones; enjoyment is about the feeling, not the outcome	Enjoyable drives are ones where you improve your limits, or win (not necessarily relaxing or pleasant)
Consequences of (the vast majority of) mistakes are insignificant / too ambiguous to notice. People get away with mistakes	Consequences of mistakes are immediately apparent (e.g. against the clock), and may be very significant (e.g. a crash)
Purpose is to complete a journey safely, comfortably and efficiently, perhaps while presenting social status	Purpose is to beat the opposition by driving faster than anybody else, and/or to improve on own past best performance
Satisficing[14] behaviour: the goal is to be 'just good enough' with a sensible investment of time and effort	Optimizing behaviour: the goal is to be as good as you possibly can, investing whatever it takes

A MOMENT IN TIME

Throttle pinned, engine pulling cleanly, screaming to the redline before each upshift… and crossing the start-finish line to clock the overall fastest lap time of the race weekend. This is how every racer would like to end every qualifying session. So, how do you get there?

Well, we could look back at all the prior events that must have taken place to lead to this. We could look a few seconds into the past, at the driver's technique in the final corner. How he is finessing the throttle and then firmly planting it to the floor, while balancing the car with slight corrections of steering. We could investigate into the neural and muscle activity during these few seconds. And there we would see billions of nerve cells and muscle fibers firing off in coordinated, synchronous patterns…

Or we could look a little bit further back in time, and also take a step back to take a wider look at how the driver has picked his 'racing line' through this bend. And we could then investigate the perception and attention and memory processes behind these decisions.

Or we could take an even broader view, and find out about his mental preparation on the day; and the problem-solving and communication that went on in the engineering debrief based on the previous day's practice sessions…

Then there's the development of social relationships with his engineering team that year, and his role in car development which has been instrumental in putting underneath him the machine he needs to do the job…

And his prior experience, the successes that led to him getting the drive in the team in the first place… ultimately, all the rainy, muddy winter weekends spent pounding around a local kart track; the sacrifices made by his family in the interest of developing his talent… all the way to his very first just-for-fun go at karting.

Or we could go even earlier: we could analyze the knowledge and skills that have accumulated within the team, and in the expert community, even long before he was born (some of which knowledge he will have absorbed).

Or even further still: all the way to the biological evolution of the visual, motor and memory systems of the brain; the millions of years of adaptation behind the sophisticated mechanisms, which in this person have been developed in a particular direction, and to an elite level of expertise.

Figure 1-1. Once inside the cockpit, the driver will be isolated from the rest of the world; there are not many sports where the athlete can so often be completely hidden! The small movements he makes are difficult to observe from a distance, or on TV. Even when you get an onboard camera view, drivers are so tightly belted into their seats that from the small movements they are able to make it is pretty tough to identify meaningful differences. To the casual observer, all this can make the skills involved look deceptively simple.

Within each of these choices, we would be looking deeper into time; to the complex chain of events leading to this singular moment, crossing the start-finish line.

We choose to start at the moment the driver is about to get into the car. This will give us a good balance between detail and an overview, and some observations we can then later build on.

So here, fifteen minutes earlier, we find him in the pit garage, in the middle of an organized bustle of activity. He puts on his helmet, then tightens the chin strap. He is not looking at the team personnel moving around him. Rather, his eyes stare in the distance. His mind is already elsewhere. Only things that make a difference to the task at hand get registered; everything else is filtered out as 'meaningless' to him now, just 'noise'.

Ten minutes to go. He walks to his racing car and steps in. Always from the left side of the car, left foot first, a habit. (Or a superstition? Is there even a difference?) As he drops down into the seat, you can tell he has done it thousands of times. The movements are economical and relaxed, but at the same time precise and purposive. Smooth, the way a hunting animal's gait is. Game face on, muscle groups working in synergy,[15] the movements are made automatically and without hesitation.

Because the movements have become an ingrained habit of the unconscious mind, the precious resource of conscious attention can remain focused, working on the overall task.

Everything around the driver is proceeding in a familiar and predictable, almost choreographed, sequence. Everybody has a job to do, and everybody knows what they are doing; without hassle or uncertainty that would be sensed by the driver. There are no surprises, no last-minute panic in anybody's movements or tone of voice. Not only the driver but also the team is in a kind of flow.[16]

In the cockpit, our driver feels his feet on the pedals and places his hands on the steering wheel momentarily, just to check that everything is in place. He glances at the mirrors, making sure that they are angled correctly. He's aligning his inner sense of space with the physical space of the car.

The seat has been moulded into shape to closely follow his body contours, and the controls are placed to ergonomically suit his size, joint movements and biophysics.[17] Cocooned, feet-up in a kind of fetal position, he is now in his own little piece of the universe. Here, for the next ten minutes, any one or any thing that is not to do with driving a racing car isn't going to intrude.

If we could now peer inside our driver's body, we could observe his physiological state. He is sitting down, relaxed in a supine position, almost lying down. And yet we would see that his heart rate is elevated from its normal well-trained athlete's resting rate. The heart responds to the tension and excitement, even when there is no physical effort and stress.[18]

The balance of blood circulation has also been already diverted away from the gut and the intestines, ready to deliver oxygen and blood sugar into the central nervous system and to the muscles.

Neural synapses in the adrenal glands on top of his two kidneys are spiking, and the glands are pumping out cortisol into the bloodstream, where it is then carried into every cell in the body.

Inside his gloves the eccrine-type sweat glands of his palms have opened small pores in the skin. This is the mechanism behind sweaty palms. (A different type of sweat gland is responsible for body temperature regulation.)[19]

These are all among physiological markers of high sympathetic autonomic nervous system activity – sometimes called the 'fight or flight' system. They signify a high state of alertness – but not the wide-eyed, overwhelming, 'deer-in-the-headlights' kind of arousal. No, this is the cool and the calm of attentional clarity and focus.

The 'stress hormones' are not creating anxiety, or swellings-up of negative thoughts and emotions. They are instead delivering a laser-like sharpness and intensity to every perception, thought and action. The brain is mobilizing energy reserves and mental resources. The body feels good, energized. His brain is registering that the body is 'pumped up', but at the same time knows that this is how it should be. So there is tension, but it is a good tension: an eagerness to 'get it on'.

There's a flurry of activity in the adjoining pit box. The sister car is coming back from the track with a flat tyre and a busted rear wing – possibly also damage to the chassis underside, its aerodynamic 'floor'. (Driver: '*F**K – the f**king ***hole cut me off!*' Race Engineer: '*Box. Box. Box. Box.*' Driver: '*F**king idiot!*' Race Engineer: '*Yeah. We saw that, we are reporting. If the floor is okay there's still time to send you out on another run. Just take it easy coming in*'.) Mechanics are running out to get a new set of tyres, others to receive the car and pull it back into the pit box. The chief designer is walking from the pit wall to the box to have a look at the chassis for himself, to eyeball the damage.

None of this matters for our driver. None of this barely *exists* for our driver. It could as well be an anthill seen from an airplane. All that matters, all that exists right now, for the next eight minutes, is a 3-dimensional layout of tarmac and the speed at which it will be taken.

Observe next, if you will, his eyes. You will see his pupils are slightly enlarged (another marker of sympathetic autonomic arousal), as if to take in as much information as possible. And his eye movements, like all the movements of his body now, have an air of economy and focus.

He's shut out the rest of the world in his mind – and so have his eyes also shut out any extraneous stimuli in the visual field. They do not flit nervously around the scene, searching for the next thing, trying to figure out what is happening. He very calmly observes only the objects and events that his mind chooses to attend to.[20]

Some of these things would be: the timing screen placed on top of his dashboard; the mechanics holding tyre blankets over the tyres, freshly brought out from the pre-heating ovens; the look of the sky and the flags on top the opposite grandstand (the cloud cover and wind direction are relevant to ambient track conditions, and are therefore duly logged into memory).

And what would we see if we were able to peer into the brain itself, and make out the myriad neural circuits recruited to action? The physical context of the racing car – the feel, the sight, the sounds, the smell of it – has immediately evoked memories, motor plans and emotions specific to race driving.

Suppose our driver has decided to use the final few minutes of waiting time to visualize in his mind the perfect lap. Eyes closed, he's running a 'mental movie' of the lap in real time.[21] This type of process – what psychologists call mental imagery – involves many of the same perceptual and motor networks in the brain as the actual performance will. So, if you were able to measure the activity of his brain, you would see signals whizzing across the visual cortex, for example. But also the motor areas of the brain would be active. Only, they are not currently sending the final 'go' signals out to the muscles.

Of course the normal sensory data – the sensations that would actually be experienced in the physical situation – is not coming through the eyes, or the other sense organs. The sequence is played out as an internal simulation 'in the mind's eye' only. What this means, physically, is that the 'mental movie' – the succession of visual and motor processes – is being generated on the basis of memory system activity, not on the basis of sensory signals.

Here the part of the brain that is calling the shots, running or rewinding the mental movie as it were, is the 'executive' front part of the brain (not the sensory-signal-analyzing part at the back of the brain). It's the evolutionarily new part, the one that enables foresight, decision making, impulse inhibition and complex action plans. It is this executive part of the brain, called the prefrontal cortex, and the deep memory networks from which past experiences are retrieved, that are driving the visual and motor systems, not input from the eyes and body.

The body is relaxed, the final motor pathways responsible for actually *activating* the muscle groups are out of the loop. But if you used sensitive enough instruments, you could pick up just some tiny twitches of the muscle fibers of the head, hands and feet. And behind the closed eyelids movement of the eyes might reveal to the outside world the progress of this process. (Within the confines of the racing car he is not able to do the full-body movements sometimes seen in pre-performance visualization in some other sports.)

'OK mate, you were losing time on the entry to T4. We think moving the brake bias forward might be a good idea'. The race engineer's instructions, radioed from the pit wall, interrupt the flow of images. Someone walks over and hands over a printout that is showing his vehicle telemetry overlaid on his teammate's, from their previous runs.

Just very simple and easy-to-use information at this final stage of preparation: graphs showing traces for speed, gear, brake and throttle pedal positions and elapsed time at each point on the track, on their respective fastest laps, with the locations of the different turns of the racetrack numbered on the horizontal axis.

He studies the graphs to see where he has been gaining and losing time, and how he and his teammate are operating their vehicles differently. This symbolic information is interpreted, becoming instantly meaningful to him within the context of his memory: all the laps he has done, his knowledge of the circuit layout, the way he knows this racecar handles on this particular day, and how it will likely respond to particular inputs whilst braking, cornering or accelerating at different speeds… In his brain resides a huge – and really, literally, astronomically huge – body of knowledge of such things.

This knowledge allows the brain to automatically predict what outcomes his actions would have. *What would happen if I left my braking a little later in turn four? Or what if I try carrying more speed?* In fact, much of the same predictive machinery is at play here that was running the visualization just a moment ago. Only now the brain is doing something perhaps even more remarkable: from the observed outcomes, such as the speed and throttle positions at different parts of a bend (actually, just symbolic representations of them on a piece of paper!) the brain is working backwards, back to the actions and the decisions that generated them. An unconscious inference leading to *'Yeah, looks like he wants to carry more speed through T4…'*[22]

He can immediately 'see' that his team-mate is braking later and turning in at a higher speed, yet getting on the throttle at the same place. Will he also try this on his final run? Would it work for him, given their differences in temperament and natural style? Would it work, given the way his car has been set up? Or is the team-mate running a higher level of downforce? (Faster through the corners, but also creating drag that hurts acceleration and top speed on the straights?) No, the aerodynamic set-up of the cars is essentially the same – this relevant fact has been recorded somewhere in the depths of the cortex during the debriefing session, and flashes through his mind so automatically it is hardly conscious at all. So, the speed difference should be down to the driving technique (and perhaps minor differences in suspension settings). He will try it. Decision made, this new information updates his mental picture of the perfect lap. He can 'see' himself taking the corner in this new way, and how the car should react.

There are numerous other paths of reasoning our driver *could* have followed up on… but did not. For example: does this new decision imply changes to the suspension settings, to make the car stable and willing to turn in at the same time? But then, will that have consequences on how other corners of the track should be taken? How important is this particular corner for overall lap time, anyway? And if a suspension setting *is* changed, is he likely to gain more here (and similar corners – are there any?) than what he will lose elsewhere?

But there is no time for such complex problem-solving now. The car balance was feeling good. The change in cornering technique in turn four either will work, or then it won't. End of. A firm decision has to be made, and then not questioned. And it's not a good idea to change many things at a time.

Those irrelevant paths of potential options do not bubble to the surface of consciousness to disrupt the state of concentration – or maybe just enough to make a mental note to bring this up in the debriefing, when comparing the setups and sector times of the two cars.

Six minutes to go to the end of the qualifying session. The signal is given to turn on the engine. A mechanic inserts the starter motor and cranks it. It catches and exquisitely balanced masses begin to rotate and reciprocate behind the driver's head. The noise bounces off the garage walls, and the buzzing vibration through the seat tells the brain that the situation has changed. Simple movements of the body will now make the engine respond. The car is 'alive'.

The race engineer is calling the time to go out; he's waiting for a gap in the 'traffic' on the track: other cars leaving for a warm-up lap, on their flying laps, or coming back to the pits. And anyway, one should go out as late as possible so the track will have rubbered in. (The material of hot, soft racing tyres deposited on the asphalt gives more and more grip on 'the line' the further the session progresses.)

Tyre warmers are taken off. The car is lowered from the jacks. The driver is now in full control of the vehicle. Clutch down, gear in. (Clunk). Pit road is clear, the signal is given to GO. Release the clutch and drive out of the pit box and down towards the pitlane exit… and then to the stretch of tarmac beyond… onto the racetrack.

Just driving up the pit road the car is already talking to the driver. The feel of the tyres, the suspension, the sound of the engine, are all telling him that everything is alright, as expected. The feedback from the eyes, ears and the skin is giving the racer's brain all the right messages, the ones the brain is already predicting. Nothing is out of the ordinary, nothing in particular catches the attention, nothing breaks the flow of concentration.

Out of the pitlane. Speed limiter off. The engine pulls as it should (organs of balance in the inner ear and pressure sensors in the skin are reporting physical acceleration). Gears engage smoothly (no hesitations in acceleration, no unexpected sounds from the gearbox). All this is more information to be logged into that huge body of knowledge in the internals of the brain: more details about how this particular car felt and responded on this particular day.

He starts building up heat to the tyre surface by fast cornering, acceleration and braking. Braking also warms up the brake discs, and the heat soak from the brakes raises the temperature of the rims and the tyre carcass. The car needs to be prepared like a tuned musical instrument, every time.

This is all routine – a habit – but also needs to be adapted to the conditions such as the grip, and the ambient temperature. Because these are not pre-packaged reflexes, the prefrontal part of the brain is once again needed, to adapt the selection of motor routines to the conditions.

Sensory information is also now being fed in: touch, sound, sight, sensing g-forces. An overall sense of how the car responds and 'what the car needs' is all getting logged from tiny cues, and integrated into an almost unconscious, subliminal 'mechanical feel'. The speed and intensity of every action is adjusted accordingly, as are the parameters of the mental model of a perfect lap.

A quick glance to check out the big leaderboard by the side of the track to see the running order. Who's got provisional pole? The engineer comes on the radio: *'How does the car balance feel? Remember to warm up the rears... The wind has picked up since the previous run, and it's now a 5-meter headwind along the main straight...'*

The wind is a relevant bit of information for the driver's brain, to further update the mental model of the upcoming lap. For one, it means you can brake later and deeper into the headwind entering turn one – but you may get into trouble with tail wind mid-corner in turn four... especially so if you are now going to be experimenting with that carrying-more-speed thing...!

Billions of nerve cells are crunching this data in the background, in preparation, to make the split-second decisions of how to handle the corner as fast and automatic as they can be, once you get there.

Warm-up lap now coming to an end. Adjust the knobs on the steering wheel to set engine and gearbox and hybrid energy harvesting/deployment for a full-out attack. Differential and brake balance in the right settings for turn one at the beginning of a quali lap. Check, check, check, check, check.

A few 'tyre warming' yanks on the steering wheel to feel the grip again. Slow down a bit to have more clear track ahead... the car in the distance is leaving for their timed run. And it's a fast one, so, with a little bit of free air between the two of you, you won't catch it until maybe the start-finish straight to get a bit of a tow from its slipstream. Can't slow down too much, though. You don't want to be caught up by anyone from behind. Take the final corner on a line that gives you maximum acceleration down the start/finish straight. The lap begins...

Let us for a moment leave our driver to do his business, and step back to sketch out some of the things that the next two minutes or so, 'doing a lap' will consist in.

First of all, you can divide up the lap into bends and straights. Many famous bends have names like Eau Rouge, Carousel, Corkscrew or

Ballagarey (also occasionally called Ballascary – *'because it is'*). More often than not, though, racing drivers and engineers just refer to the corners by successive numbers. Turn one, Turn two etc. Each bend can then be divided into distinct phases: braking, turn-in, entry, cornering/apex and exit.

Already before braking for a bend, the racing driver will have had to anticipate the precise trajectory they want to take through the corner: the racing line. The classic racing line[23] is 'out-in-out': starting from the edge of the track (on the left for a right-hand bend, on the right for a left-hand bend), cutting to the inside to 'clip the apex' of the corner, and then accelerating out to the edge on the exit. This will round off the corner, increasing the radius of the car's path, and allowing a higher speed for any given amount of lateral g the tyres' grip can generate.[24]

What is not quite obvious, unless you are a racing driver or a race engineer, is that taking the racing line is not the same thing as 'cutting the corner'. Cutting too early, and cutting in too much too soon, actually takes you way off the racing line. This is in fact a typical textbook error[25] for beginners: as you are going into a bend at a speed that seems higher than what is perhaps healthy, your brain's survival instincts kick in, and there's an urge to start steering, to avoid the bend rushing towards you… But this means that you arrive at the apex too soon, and with the car pointing in the wrong direction – too much to the outside of the bend. And so, in the end you need to slow down too much to make the car rotate. (In driving technique terms, you are 'apexing too early'. The proper racing line, in contrast, turns in later, rotates the car more on the entry, and then takes something called 'a late apex', allowing a shallower arc on the exit, and therefore better acceleration.)[26]

When driving fast, it is not only the geometry of the racing line that the racer's brain needs to concern itself with. How the car's weight is supported by each of the four tyres, and how this vertical load shifts around with every input the driver makes, is crucial. This is something you do not really physically experience in everyday driving, nor, with modern cars especially, do you see it very clearly on the TV – at least not if you don't know exactly what to look for.

There are just four contact patches – the fairly small surfaces where the tyre grips the track and through which all acceleration, braking and steering is achieved. How much grip a tyre is able to give depends quite critically on the load that this tyre is under. And throughout each turn, the driver is modulating the front-aft and side-to-side distribution of tyre loads, with inputs from the brakes, the throttle and the steering wheel.

If you look at these things from an engineer's point of view, there is a maximum force, the combination of grip and traction, that a tyre can produce. This is approximately the same in every direction; a fact that is often depicted by a concept called the friction circle.[27] When all of the grip the tyre can generate is 'used up' for cornering (steering), there is no traction left for accelerating or braking. Likewise, if everything is used up for acceleration or braking (you have wheelspin or a skidding tyre) then there is none available left for steering. The skilful racing driver will at all times try to live as much as possible on the edges of the friction circles. But more than that, because the magnitude of the grip and traction available from a tyre, the *size* of the friction circle, depends on how much load a tyre has on it, the driver is all the time also managing this, for each corner, through moving the total load around between the tyres.

For an expert driver all this dynamic complexity is known in a largely unconscious manner. With years of practice, the process of taking into account these dependencies in one's driving technique becomes highly automatic and intuitive. Little things a driver will 'pick up' when becoming familiar with a car, and a race track.

Using this knowledge does not happen in words and technical terms, such as you would write in a book. Instead, all this knowledge (and more, much more, this was just the simplest 'dynamic load transfer for dummies' account)[28] is stored in the form of what psychologists call implicit procedural knowledge. Implicit knowledge means things you know without realizing you know them. Procedural knowledge means knowledge that is only expressed through action; knowing what to do, and how to do it.

So, the driver is not so much 'thinking' his way around the track, in terms of engineering concepts, but more like 'feeling' his way, in terms of whatever underlying concepts a racing driver's brain develops when 'picking up' this deep, unconscious knowledge base.

Let us explain this with an analogy. (We'll try to use a lot of analogies to help you along.) It's like a chef knowing that certain ingredients when blended in a particular way 'go well together'. They can't necessarily describe the exact flavours in analytic detail – let alone the *chemical* properties of the molecules that produce the sensations! It's implicit, experiential knowledge. You know it when you feel it. And they cannot describe in words all the nuances of technique either – if they want to teach you how to prepare the dish then they'll have to *show* you, not just *tell* you how to do it. It's procedural knowledge and cannot be fully put into words.

We can take this analogy a little bit further still (and it remains valid): just as 'taste' does not quite capture what is going on when you enjoy a well-prepared dish, so also the word 'steering' alone does not quite capture what is going on when a driver is guiding the car around a bend.

A chef will not be concerned merely with 'taste' because the sensation of smell is integral to flavor (the tongue only distinguishes, quite crudely, between sweet, sour, bitter, salty and umami). And also texture, mouth-feel, temperature, moisture and numerous other sensory features are subtle but very essential characteristics of a dish; and feedback to the chef about whether their recipe works or not.

Likewise, talking with drivers, you will hear about subtle but very essential characteristics of steering: 'loading' the chassis or tyres before putting on more steering wheel angle, or 'balancing', or just 'cornering' – rather than 'steering left or right'! And in the middle phase of the bend, especially, you may have very little steering wheel input, you are mostly steering with the throttle.[29]

So, the racing driver puts together a cornering technique by blending different control actions and 'feel' much like a chef puts together a dish by blending flavours and 'feel'.

This brings us back to our driver, who will now have executed this 'blending' some 10 to 15 times (the number of corners on a typical modern racetrack). He's now 'steering with the throttle', out of the final corner – it's a fast, right-hand sweeper – and our driver is fighting with the car just a little. According to Mario Andretti, *'if everything seems to be under control you are not going fast enough'.* [30]

His body is taking a pounding and subjected to heat, noise, vibration, and g-forces. At the same time, he's delicately balancing forward thrust against lateral acceleration, at the edge of available grip. Full power now, shifting just at the right revs, going up the gears – bang, bang, bang! – and across the start/finish line.

Return to normal reality, and immediate feedback: *What's the lap time? What's my position? I was fast... but how fast was I?*

Chapter Two

Analyzing the Complex Skill

THERE IS COMPLEXITY involved in driving a lap. In fact, even the skills that are needed to just take a car around a single bend, on the racing line, at racing speeds, are many. If we now tried to describe everything at once, we would not be able to get started.

Take any one moment in time – let's say the moment when the driver is turning into the bend. Now, try to observe and try to describe *all* the skills and *all* the techniques that the driver must master. All the subtle nuances of skill, all the myriad factors in vehicle dynamics which the driver must be sensitive to because they will determine failure or success. Try to recount the innumerable underlying brain processes that get expressed within that apparently simple action of turning the wheel. Now put all this into the context of the design of every component of the vehicle, which together produce those dynamics and sensations that lead to those specific actions…

We could fill an entire chapter (or an entire book) with material on this single snapshot. It might even be an amusing read, at least for someone already familiar with the subtleties. But it would be impenetrable for many and, more to the point, it would not directly connect with knowledge from cognitive science and brain research (except perhaps here and there).

So, instead we must now begin a process of breaking down the complex performance, back into its elementary parts. Working towards the question: what problems the does the brain have to solve, in order for you to go fast?

What we are going to do here is to start with the foundations: a proper analysis of how you would go about answering such a question. Then, once we are more clear on what the problems are, the details of how the brain goes about solving those problems will fall into place.

Please bear with us if this chapter is slightly more theoretical; different alternative ways of breaking down and analyzing driver performance are really different theories of driver performance. Different viewpoints you could take: the engineering, the ergonomics, the cognitive science perspective. Each viewpoint presents a partial view; from these 'the whole picture' will begin to emerge.

Before we plunge in, though, we will tell you up front what particular framework we will eventually choose[31 32]: it classifies the skills and cognitive processes into a hierarchy of levels, called the control level, the guidance level and the navigation level. For readers familiar with vehicle systems dynamics, and the concepts and techniques used in driver-in-the-loop simulation modelling, this will be a familiar classification and may require no further justification. But for others, we want to use this chapter to explain *why* we think it is a particularly good way of looking at things.

(And also how it relates to perhaps more familiar, or more intuitive, ways of categorizing driving behaviours. As a bonus, explaining why we did *not* choose those other frameworks allows us to already start defining ideas. Defining means literally: setting of boundaries).

THE ERGONOMICS VIEWPOINT

Now, if our main concern was racecar design – or more specifically, say, the design of the cockpit interface – we might group skills in terms of operating the individual controls. We would ask: What are the basic actions and abilities involved in using the *steering wheel* (confident turn in, deft counter-steering etc.)? What is needed for modulating the *brakes* (adjusting the brake pressure to reduce speed just enough, and at the same time managing the front-rear weight balance)? What are the correct and incorrect ways to get on the throttle? And so on. This is what we would do if we were approaching things mainly from an *ergonomics* or *vehicle engineering* point of view.

It is a striking feature of motorsport – and indeed a very useful one for scientific analysis – that experts in this field display all their skill and all their expertise through such a small number of control operations.[33] And one can indeed speak of some drivers being 'good on the brakes', some having a great feel for 'balancing' the car during cornering, or having highly 'sensitive throttle control' etc. So, we could have one chapter dedicated to braking, one chapter to steering, another for throttle and so on. This would be a completely sensible classification, and would help the reader ground the new material because it builds on concrete objects and actions she knows about. Great.

However, for us the choice of organizing ideas must be guided by our ultimate need: to look at driving skill from the point of view of the different jobs that specific circuits in the brain have to do. We're interested in the design of the racer's brain, not the design of the racing car.

And when you look closely at the problem of going faster, and formulate it as one of choosing actions – making micro-decisions on how to use the controls – then almost all the actions will involve almost all the controls, simultaneously. For example, corner entry technique involves adding more steering wheel rotation whilst reducing pressure on the brake pedal. It is crucial that these are at all times in the right proportion *relative* to one another.

Technically, what one learns when learning to properly control a vehicle is to master, and to take advantage of, the nonlinear interactions, dependencies and relations between the controls. How hard one is braking affects the distribution of weight on the wheels, and therefore how hard one can turn – how much steering lock the car is going to 'accept'. Similarly, how hard one is cornering affects load distribution, and thereby the acceleration the vehicle can produce – how much throttle the car can 'take' in that particular dynamical state.

For the brain, this means that it must learn motor programs for car-control that take into consideration all these complex connections between the controllers. (More on motor programs in **Chapter 6,** stay tuned). Thus, the organization of *skills* does not, from a psychological point of view, separate to discrete skills with individual controllers.

And the brain definitely does not have a 'throttle centre' or a 'brake centre' (or any you-can-name-it-easily-centres at all, for that matter). There is, for sure, specific anatomic connectivity from certain parts of the brain's motor areas to the foot, say. So, those parts of the brain are of course more directly connected to the throttle pedal; from other parts of the brain's motor system other nerves connect to the hand, and so to the steering wheel. Yet consider the way the racing driver needs to physically *move* her hands and her feet *together*, and predict what responses this will generate from the vehicle. Pushing more with your hand, say, can have radically different outcomes in different situations. (Let alone in different types of vehicles, especially if you compare racing cars and motorcycles!)

Also: a lot of the skill in 'reading the racetrack' and 'picking a line' are more or less the same regardless of the type of vehicle one is using. You don't suddenly become completely 'lost' when you step in a new kind of vehicle. So, this skill has to depend on other types of perceptual-cognitive brain functions and not just motor coordination of individual limbs, less so individual controls. There has to be a of stuff lot going on at this higher level of abstraction, which is the same in different machines, independent of the particular control interface. (One example would be the visual strategies a racing driver uses for *guidance*, **Chapter 4** tells you about these.)

Abstracting away from the ergonomics of individual controllers allows us to discuss some of these important and interesting skills and techniques, which are the same in different vehicles. I (OL) like to think of many them as 'recyclable' skills and abilities that can be transferred across different environments. Or even a kind of 'timeless essence of racing driving', not restricted to a specific technology.

THE STEERING VS. SPEED CONTROL VIEWPOINT

Another option could be to consider each action in terms of how it is related to *steering* vs. *speed*. Or: *lateral control* vs. *longitudinal control*, in engineering jargon. (Lateral and longitudinal mean 'sideways' and 'backwards-and-forwards', respectively.)

We could divide the perceptual and motor control demands, and then the brain processes, into those for braking and throttle control on the one hand – longitudinal control – and those for steering on the other hand – lateral control. Many theories and driver models[34] make just such a division. They treat lateral and longitudinal control as separate 'phenomena', or separate classes of problem.[35]

Psychological models tend to be either 'steering models' assuming constant speed, or 'speed control models' assuming the vehicle is following a set path. And the way psychologists and brain scientists then set up their experiments on steering often make the participant drive a simulator that is set to chug along at a constant speed, so that they can purportedly study 'pure steering' before moving on to the more complex task of how steering and speed are put together.

So, if we went down this route, we'd have one chapter dedicated to speed perception, another on motor programs for braking and acceleration, then perhaps one on how humans perceive and control their direction of locomotion (i.e. steer), and so on. Brilliant.

However, the racer's brain does not have this luxury. Much of the true skill of the racing driver involves keeping all four tyres as close to the edges of their friction circles as possible (see **Figure 2-1**, overleaf), and the sizes of the friction circles as large and usually as even as possible. As the size of the circle depends on the vertical load on the tyre, managing the transfer of loads from one tyre to another and back is crucial.

And here's the thing: turning the *steering wheel* produces substantial weight transfer (it will also slow you down due to tyre scrub). *Brake and throttle* use does the same, and will also steer the vehicle. Because of dynamic load transfers, speed and direction are intertwined: any steering action will slow down the vehicle; any braking or acceleration on a curved path will steer the vehicle, and whilst on a straight path, braking and acceleration limit the maximum steering you can initiate.[36] These nonlinear interactions between lateral and longitudinal control are the basic thing that the racer's brain must master.

Therefore, in the driver's mind/brain these are probably 'mixed up', blended, in a way that it takes an engineering degree and physics to *unmix*. But we are primarily interested in analyzing how the driver's brain sees the world, not the physics of vehicle behaviour, as the engineer sees it.

So, one cannot really separate the expert driver's skill into lateral control skills and longitudinal control skills. Except, perhaps, for the purpose of making a simplified lab experiment, which thereby deliberately ignores just about most of the skill! The question of how lateral and longitudinal control come together is really the fundamental one.

And one is not likely to find a 'steering cortex' and a 'speed cortex'. Counterintuitively, one has even *less* of a chance here than looking for 'throttle pedal cortex' and 'steering wheel cortex' because the controls at least do map to body parts, which then map onto the brain.

There are studies, though, where participants perform 'pure' steering tasks in a brain scanner, and they do show consistently certain brain areas lighting up. But that does not mean you have found 'the steering cortex'. (What *does* it mean, then? All will be revealed in time – this particular thing will be revealed in **Chapter 7**.)

THE RACING LINE VIEWPOINT

Okay, yet another way to organize our thinking could be to look at the different cornering phases: the basic building blocks of 'the racing line'. We looked at this a little bit already. But because this is such an important concept, we want to expand on it here.

The way that a corner is taken in a racing car forms a fairly fixed, stereotypical sequence. Approach to the bend, the application of the brakes and then changing gears, the turning of the steering wheel on entry, then balancing the car mid-corner, and finally accelerating out. (**Table 2.1.** and **Figure 2-2**, overleaf.)

We could use this temporal order, and the actions that take you from one phase to the other as the framework. So, how about settling for a chapter on braking, another chapter on cornering and finishing with a chapter on exit? Let's get on with it, huh?

Hold on. From the point of view of the brain's job description, what one actually has to get done in braking will depend a great deal on what needs to happen later in the sequence.[37] So, you need to understand these later parts before you can understand the earlier ones. One brakes *in order to* achieve the speed and load distribution that allows turn-in to happen. Yet how turn-in happens is dictated by how one has braked. So, you need to understand the earlier parts before you can understand the later ones. Braking cannot be understood except in the context of entering, and entering cannot be understood except in the context of braking...

And also: braking and entering is just to set you up for cornering, which itself is just setting up for the exit. So, you need to look at the exit first. Only you cannot, because what happens at the exit can only be understood as the consequences of what you did earlier on... hmm...

To see all this more clearly, we need to look in a bit more detail at how the racing line is put together. The most important geometrical property of a bend is its radius. Or rather: the radius of the racing line it allows you to take. This makes a big difference to how fast the bend can be taken.

The racing line itself is a path that creates a larger radius for the car's trajectory. Larger, that is, than simply driving parallel to the edges of the track, e.g. in the center (refer here to **Figure 2-2**, overleaf).

This line allows higher speed at whatever centripetal acceleration

Figure 2-1. A tyre generates forces acting on the vehicle at the contact patch; the magnitude of the force that can be produced is proportional to the load on the tyre. It is common to analyze vehicle motions separately into longitudinal motion (lon: in the direction of travel, braking and acceleration) and lateral motion (lat: 90 degrees to the direction of travel, steering). A tyre will be able to generate some amount of force longitudinally, and some amount of force laterally. A tyre can't brake or accelerate when producing maximal lateral force, nor can it steer when spinning or skidding. The maximal *combined* force in each direction yields the friction circle of the tyre. Here the concept is illustrated with the left rear tyre of Daniel Ricciardo's Red Bull at Monaco.

TABLE 2.1. THE CORNERING SEQUENCE

Braking	**Entry**	**Cornering**	**Exit**
Shut throttle	Steer into the bend	Neutral throttle	Begin acceleration
I	I	I	I
Maximum braking	(Trail braking, shifting down)	Maximum steering i.e. yaw rotation	Unwinding steering
I	I	I	I
On-limit braking	(Counter-steer)	(Counter-steer)	(Counter-steer)
I	I	I	I
(Shifting down)	Throttle 'cracked open'	Clip the apex on the inside edge	Full-throttle & track-out to the outside edge

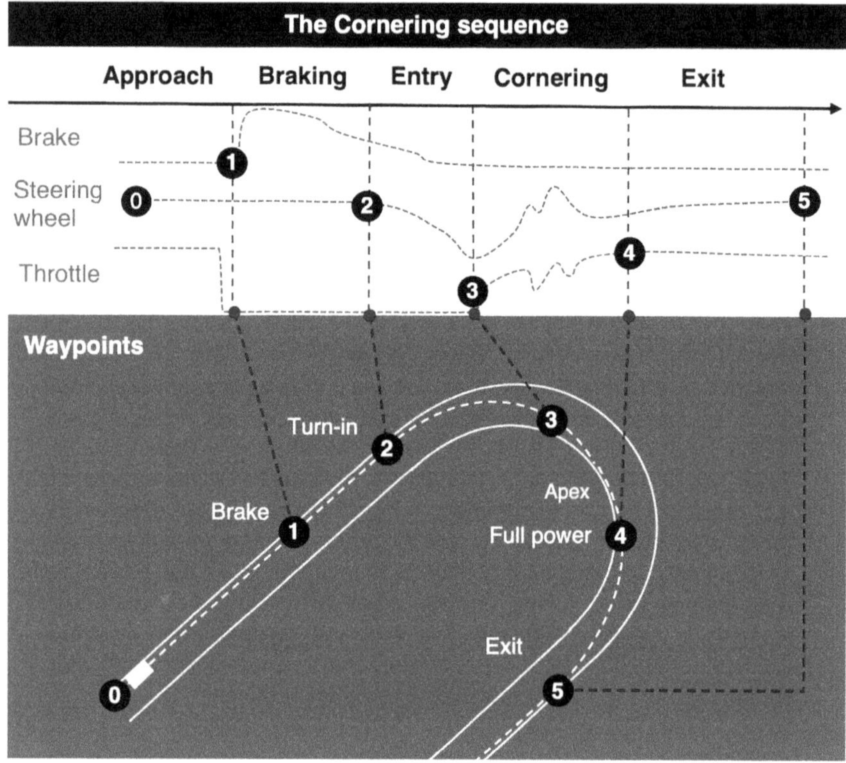

Figure 2-2. In simple terms, taking a corner can be broken down into a logical sequence of successive 'steps'. This describes the action as discrete phases that can be identified by *when* different control actions are taken by the driver (cf. the telemetry graph sketches at the top). The geometry of the racing line relative to the geometry of the bend describes a continuous path. This can be sequenced by projecting the phase-defining action points into space as 'waypoints' so that the racing line 'connects the dots' (bottom).

– lateral grip – the tyres are able to generate. This is of course conducive to the overall goal of minimizing elapsed time.

The trajectory is not symmetrical, note how the path curvature is not constant from the beginning of the turn to the end and the apex point is located beyond 'the geometrical apex' of the bend. (At least in this standard technique, the 'late apex'.)

Sequencing the racing driver's actions into what happens during approach, braking, turning in, entering, cornering, apexing and exiting identifies locations on the track where a particular control action is performed. We will call these locations *waypoints*, borrowing the term from piloting and navigation.

There is, for example, the location at which the brakes are applied, this is the braking point; the location where the steering wheel is first turned, the turn-in point; the point where the car touches the apex, the clipping point etc.

One can metaphorically think of these waypoints, connected by the vehicle's path, as forming 'the racing line', analogous to the way a child draws a pattern in a connect-the-dots picture. Of course, such an analogy is a simplification in many ways. It is a simplification, firstly, in terms of the number of waypoints. A skilled driver will perform much more than four actions in a bend! It is also a simplification because it looks at the racing line as if the actions were performed instantaneously (at a single point, rather than a location or a zone or a mini-phase, over time). And it's also a simplification in terms of describing the line as a one-dimensional path: a physical car has a finite size, and the contact patches of the four tyres do not follow the same geometric path – the skilled driver will 'place' the inside front wheel at the apex and the outside rear on the exit. But as a rough first approximation, it is useful.[38]

An expert racing driver will also consider the details of the road geometry as well, not just the overall shape of the bend. Changes in radius, elevation and camber (longitudinal and lateral inclines), bumpiness, changes in grip from different road surfaces in different parts of the bend… all these affect the choice of how to adapt the racing line to the situation: where and how the driver will turn in, where she will cut closest to the inside, and where she will exit the bend etc.

These more complex and nuanced details in how racing lines are 'stitched together' from waypoints can be seen as variations of the same basic theme, a bit like melodies and compositions are 'stitched together' from notes and phrases (another useful analogy we'll return to).

The basic theme goes like this… The braking zone begins on arriving at a braking point. This is the first waypoint, where the driver gets off the throttle and on the brakes. The braking point is wherever the driver chooses, on this particular lap, to apply pressure on the brake pedal. (Braking is not an all-or-nothing affair so in actual fact brakes are applied over some zone of finite span of time and space; but like we said, let that rest for now).

Where this happens is an intuitive judgement, based on *spatial memory* for the bend's geometry, on the visual preview of the approaching bend, and a sensitive estimation of speed. On a familiar racetrack, the location of the braking point will be more or less pre-determined. It is not uncommon for drivers to explicitly identity brake markers: specific landmark locations or objects, such as a patch-up in the asphalt or a tree at the side of the track.

When driving on an unfamiliar road, on the other hand, previous experience, the visual appearance of road curvature ahead, and the driver's knowledge of the dynamic capabilities of the car will suggest an entry speed, and an entry point; these combine with an estimation of approach speed to determine the judgement of braking distance required.[39]

After an initial hard application of the brakes, the racing driver will modulate the braking force, progressively bleeding off the brakes. (Note that this is the opposite to the habits of many in everyday driving, where an initial soft application of brakes is followed by increasing the brake pressure). Brakes are modulated to slow down to reach the desired turn-in speed at the turn-in point, which is where she chooses to steer into the bend.

Gears are shifted, at least in the classical techniques, between the braking point and the turn-in point, once the car has slowed down to a speed appropriate for the next lower gear. With modern technology this can be postponed well into the bend. From the point of view of motor actions, changing gear could be as simple as pulling a steering wheel paddle with one finger… or it may require a little dance called 'double de-clutching', something of a lost art in everyday driving but a core skill in the expert driver's skill set. (And for some strange reason a *very* satisfying way to connect with the vehicle.)[40]

Braking has the effect of reducing the speed of travel, obviously. But crucially, it also shifts the load (weight) of the car, so that front tyres carry more of it than the rears.[41] The desired entry speed should be achieved at the entry point location, at which point the car should be in the correct gear, and it should *also* have an appropriate amount of load on the front wheels. Only when all these conditions are met, at the same moment in time, *that's* when you are supposed to turn the steering wheel.

Where the driver gets off the throttle and onto the brakes, how hard this is done, and the rate at which you ease off the brakes, all depend on where, and at what speed, and in what state of dynamic load distribution between the front and rear axles the turn-in needs to be performed. They all must fit together!

Dynamic load transfer, which is how much of the car's mass is supported by each wheel and how this changes over time, is something the everyday driver is rather oblivious to... But the racing driver will be extremely sensitive to it. It's the basis of 'feel' for the car... the *balance* of the car.

How much the car will actually respond to any steering wheel input at turn-in depends critically on the load distribution across the four tyres. To be specific, more load on the fronts makes the car turn in more eagerly but might also make the rear unstable; less load on the fronts makes the car more sluggish to turn. By modulating the brake pressure the driver is therefore not only setting speed: the load distribution is being modulated as well, in order to get the correct speed and tyre loads to coincide at the entry point.[42]

Upon arriving at the entry point, a fast and deliberate – compared to everyday driving – steering wheel action is made. From the point of view of the car, this generates slip at the four contact patches. This starts rotating the vehicle and generates the centripetal acceleration that makes the trajectory curve into the bend. Technically, *slip* is not to be confused with inducing a *skid* where rear wheels just slide without steering the car (even though it often is, even in specialist literature).[43]

From the turn-in point at the outer edge of the track the vehicle is then taken to the inside edge, to the apex, sometimes called the clipping point. This waypoint is the location where the driver places the car closest to the inside edge of the racetrack. This is also where the car will be cornering hardest, achieving maximum lateral acceleration. In the simplest case, a very long bend with a constant radius, the vehicle may be maintained for some time in a stable state of neutral throttle, constant speed and maximal lateral acceleration, 'pure' constant-radius cornering or *drift.*

Technically, a drift is not to be confused with a skid, either (even though it often is). But don't worry if you can't tell slip from skid from drift from an explicit 'physical definition' point of view. Most people can't, including most drivers – theoretically, that is. You will most definitely know the difference and can tell them apart when you *experience* them! This is one of many examples where the explicit, declarative 'knowing-that' can be quite detached from the procedural, implicit, 'knowing-how'.[44]

The apex must be taken correctly. Not only in terms of where the driver clips the apex, but also what speed the car should have at the apex, what direction it should be pointing to when it gets to the clipping point, and how the front-rear loads are distributed. So, setting up the car for the corner during braking and entry is certainly not just a question of slamming on the brakes and then 'cutting into the corner', hoping for the best…

After the apex, begins the exit phase. Here the driver feeds in the power and unwinds steering, allowing the car to track out to the exit point. Here we mean 'unwinds steering' in the sense that the car's trajectory curvature is reducing – the direction of travel of the car is 'turning less'. Where the car's nose is pointing, and what angle the steering wheel and tyres are at, is another matter. Depending on the handling characteristics of the vehicle there might be some chassis yaw slip, and lots of sawing of the steering wheel – i.e. countersteering – going on when 'unwinding'.[45]

Throttle is also increased until the point where the car can take full throttle. Ideally, once the throttle is cracked open, power is applied progressively. Getting on the throttle too hard too soon can lead to spinning the rear wheels (a rear wheel skid: oversteer). Or it can lead to a condition of upsetting the vehicle's balance (the front washing wide, off the desired line, because load transfer to the rear axle reduces front tyres' grip: understeer).

Backing off the throttle, slowing down too much, then getting on again, is not recommended – this will delay acceleration and may even lead to a severe rear-end skid as the weight shifts from rear to front. The throttle application needs to be 'smooth'. That said, 'smooth' does not imply *no* throttle modulation at all. The sensitive driver may be making small, deliberate adjustments to speed, balance and line through quick, small and accurate throttle applications matched to delicate steering inputs (more on the ubiquitous but hard-to-define concept of 'smooth' in **Chapter 5**).

The vehicle will now hopefully arrive at the outer track edge, accelerating at the limit of traction, but not run off the track. The end goal of the whole sequence is to end up at the exit point aligned with the following straight, steering straight, at full throttle and a high exit speed – and ultimately in the minimum elapsed time.[46]

Generally, the apex or clipping point must be chosen in such a way that the driver can apply full throttle before they hit the exit point on the outer edge of the track. And the turn-in point must be chosen so that you can get to the apex. And the braking point must be chosen so that you can turn in. In this way, the exit point actually determines the apex, the apex the turn-in and the turn-in the braking. So, while we can look at the sequence front-to-back – i.e. the direction of time, the direction of causes and their effects, the direction of actions and their outcomes… – the logic of *goal-dependencies* that the driver's brain must handle runs in the reverse direction. Deciding what to do runs back-to-front!

The corner presents the driver a problem. A kind of logical puzzle to be solved. You need to get to the other side in the shortest elapsed time, highest exit speed, all within the constraints of the geometry of the bend and the dynamics of your vehicle. It is a bit like solving a system of equations: what values of unknown variables must be chosen that will yield a correct solution for *all* the constraints?

Only it's different in that the problem-solving is unconscious – it's not a step-by-step procedure like its verbal description… and it needs to be done *right now,* and then you need to commit to the solution at speeds up to 200+ mph…

But in some sense even before you apply the brakes you need to know the end-result. Once you are committed to taking a bend in a certain way, there is only room for very minor adjustments and finessing the details of the line – a car teetering at the edge of adhesion will not accept any major changes of plan mid-corner.

The upshot of all this, with regard to our interest in analyzing driver skill, is that we could trace the temporal sequence: we would dedicate one chapter to reading the bend, another chapter to controlling the brakes, then have one chapter on turn-in, another on balancing the vehicle on the edge, and cap it off with a chapter on maximizing drive on the exit and what actions over what controllers need to be combined in what way. And, indeed, many driving instruction books take this approach. Or we could trace the logical sequence instead and start with the exit; we could ask: 'how does the driver get there?' and then work our way back through the 'dots' in the reverse direction.

Still, when we'd get to the brain's point of view, we would be stumped in trying to map this classification to specific parts or functions of the brain. Where are the 'corner approach circuits'? Where's the brain region that 'lights up at corner exit'?

Most processes are not specific to the approach phase or to the cornering phase or exit phase. For example, the driver's brain needs to accurately sense speed, judge distances, and predict the vehicle's response, and feel the changes in speed rotation. It needs to constantly perceive and mentally update one's location on the track... All this is something that is needed in *all phases*. So, these basic jobs the brain of the racing driver has to do will cut across all phases; indeed, they will cut across all the preceding classifications...

'Okay, fine. So just tell me what you think would then be a natural, 'brain-compatible' job description for the racer's brain', you say? Well, we're glad you asked.

NAVIGATION, GUIDANCE AND CONTROL AS A JOB DESCRIPTION FOR THE RACER'S BRAIN

We will go here with a taxonomy proposed in the 1970s[47] that analyzes driving behaviour to three hierarchical levels: *Navigation, Guidance* and *Control* (**Figure 2-3** and **Table 2.2.**, overleaf; these three levels will be the focus of the next three chapters, respectively).

Control refers to the split-second decisions the racing driver, or her brain rather, needs to make: e.g. knowing exactly how hard to brake to achieve the limit of grip. How much and when to turn the steering wheel, and how to do it to achieve a given rate of rotation.

It deals with the here-and-now. It is about what is happening inside your body, at the contact points where your body and your mind interface with the car, and with the contact surfaces between the car and the external world.

The control level is also behind the deftness and skill experts display in operating the controls; the dexterity that comes only from experience. This level is most closely related to describing some drivers as having 'great car control' or great 'feel' for the grip in the wet, for example. It's also about what you need to 'stay on top of the car' when it is 'out of shape'. It is the control level processes that spectators are admiring when they are wowed by the 'reflexes' that get a driver out of a 'situation'. (But as we'll see, in terms of neural processing what is happening is much more sophisticated than reflex.)

At the control level, we will describe sensations flowing into the brain from all its sensory input channels, the tuning of perceptual pro-

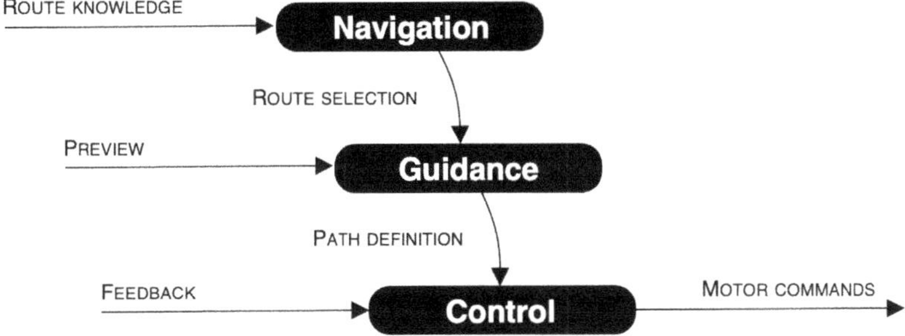

Figure 2-3. The McRuer et al. (1977) hierarchy. Navigation refers to selection of a route: which of many possible alternative paths (from your current position to a goal destination), best satisfies your needs? In racing, the basic need is the need for speed; or more accurately, to minimize the elapsed time, given some acceptable level of risk. Guidance relates the navigation-level route to what you are currently perceiving ('defining' the desired path in the visual field); it is also important for guiding timing and the 'rhythm' of sequences of actions. Control realizes the desired path, grounded in feedback based on what you are currently sensing through your eyes, ears and muscle sense. At this level the expert racer must have superb sensitivity for speed and tyre grip, 'feel' for brake and throttle modulation, as well as a practiced, confident smoothness, precision and speed of motor actions.

cesses in touch, hearing, and vision that give a driver's brain the sensitivity needed to read and interpret subtle cues in how the car is moving under you. Control level processes turn mere sensations into meaningful *feedback*.

A racing car cannot be driven at speed based on feel or a 'touch' for the controls alone, though. You obviously need to see where you are going. In terms of the informational inputs, the control level refers to locally available information: signals the brain receives about the pressures on the body, the tone of the muscles, movements of the limbs and the joints etc.

But the control level processes need to be guided by processes that are able to determine the path you need to take, based on more distant,

TABLE 2.2. NAVIGATION, GUIDANCE AND CONTROL SKILLS

Navigation	Updating long-term memory (cognitive maps) with waypoints and landmarks for reference points.
	Determining waypoints (e.g. 'figuring out the apex' based on track knowledge from long-term memory).
	Probing reference-point spatial memory with mental imagery; thinking about them explicitly, using words, diagrams, maps etc.
Guidance	Visual strategy. i. looking far ahead ii. recognizing reference point landmarks, determining direction and distance to waypoints iii. using peripheral vision to track multiple targets in 'widescreen' peripheral vision (executive control of covert visual attention to attend locations that you are not looking directly at).
	Motor imagery and visualization.
Control	'Feel' for grip & traction in limit handling. Using multisensory feedback to sense changes in friction and load on each tyre.
	Estimating speed and acceleration in linear and rotational motions.
	'Smooth' control. Selecting motor routines to adjust speed and steering, fine-tuning them to vehicle dynamic response (especially higher time-derivatives of vehicle or controller position, i.e. jerk, snap).

non-local, information sources: *preview* of the track layout at the scale of tens to hundreds of meters.

Vision will give the brain enough time (up to a second or two) not only respond and react to what the vehicle is doing now, but to actively guide and cajole it to do what you want it to do. The guidance level has to do with the special visual strategies racing drivers use in 'looking ahead' on the racetrack, and the special ways racing drivers have of using peripheral vision. These are different compared to most everyday forms of locomotion, which is necessary because of the speeds involved (stay tuned for these in **Chapter 4**). Also belonging to the guidance level are the completely internal processes for 'visualizing' the racing in your mental imagery from memory.

Navigation, finally, refers to knowledge of the track layout at a larger scale, including knowing parts of the track that are not currently visible from the current point of vantage. It is the really intimate knowledge of the racetrack that is the foundation for other skills and performance.

(This is something that may elude the casual observer – but psychological research has shown that this is the case in any complex domain, and the sheer amount of knowledge experts carry around in their heads is stupidly large.)

This knowledge is of course stored in the brain. It is memory engrained into its fine structure and remains there indefinitely – even if most of the time you are not aware of any of it... This point bears commenting on, briefly, before we go further. We have been and are soon going to be talking a lot about the 'knowledge' of the racing driver. We have taken pains to mention how most of this knowledge is intuitive and unconscious. The reason why this is important is that this sort of intuitive and unconscious knowledge *you cannot necessarily find out simply by asking a person.* (If you could, it wouldn't be unconscious!)

In everyday life, navigation-level decisions would be: *Shall I take the train or a plane? Do I take the motorway or the backroad? How do I get from the sofa to the kitchen?* Some of us are better at describing in words the layout of our house, block or hometown, others less so – but all of us are able to find our way. So, we must all have this knowledge, whether or not we can put it to words.

Similarly, some drivers may be better at *describing* a track than others, but that is not really what we are talking about here. Navigation level processes are involved in, for example, 'picking a line' through a bend. That is: determining the correct locations for the braking point or apex[48]. Whether or not you can say why you did it, is another matter.

All skilled drivers must carry a lot of track knowledge inside their brains. Even if much of it, perhaps the vast majority, is used automatically and unconsciously. This is knowledge that a person will *have,* and will be able to *use*, but which they cannot necessarily *express* in words.

What you know, and what you can do, depends on the past experience your brain has stored away and is able bring to bear on a situation. Whether or not this knowledge can be put into words – or how much of it can be put into words, rather – will depend also on your vocabulary, and whether or not you have a way with words. This is a separate skill.

So, it is not uncommon for a driver not to be able to say at all clearly what she is doing in the car – or indeed why she is fast on some circuits and less so on others. Also, when words aren't quite up to the task of expressing the knowledge, racing drivers will often use hand movements and various 'sound effects' to describe to their engineers what's going on with the car.

Making some of that implicit knowledge explicit, describing some of that knowledge which is normally engrained into the physical structure of the brain by using instead symbols on a page is what this book is about.

‘YOU CAN GET HOPELESSLY LOST BY CONTINUING TO 'LOOK' AT A TURN FROM BEGINNING TO END INSTEAD OF FROM THE END TO THE BEGINNING.’

- KEITH CODE [49]

PART II

COGNITION

Chapter Three

The Power of Knowledge (Navigation)

FIVE LAPS INTO THE FIRST PRACTICE SESSION, and the team are worried. All that potential shown in his debut year… maybe gone to waste. Maybe he isn't the real deal after all? For sure, Monaco is the toughest track on the calendar, especially with the monstrous power of an 80s turbo-era Formula One car under your right foot… but surely, *surely*, there must be something wrong?

Then comes the sixth lap, and, while the team are trying to understand what is happening, Ayrton Senna pops the car into the top of the time sheets… and keeps going faster… and faster… and faster...

'What on earth were you doing?!' the team manager later asks him. Why was he pootling around at a snail's pace? *'Learning the circuit'*, was his calm reply.[50]

Of the many challenges a racer's brain has to tackle, perhaps the most widely underappreciated is knowing the geometrical layout of a race circuit. It seems so simple at face value: you go there and you have a look. You can see where the track goes, can't you? But when you delve into it, you find it is not simple at all – it is actually one of the more remarkable feats of long-term memory performance in sports.

It is really the detailed memory of the circuit layout that affords the racing driver the foresight and the anticipation they need. This is what gives the driver the confidence to push a car towards its limits. Remember: what frightened Moss about the Mille Miglia was not being able to learn the circuit.

Not for nothing did Moss and Jenkinson reconnaissance for over ten thousand miles – they were gathering intelligence and preparing a brain-external knowledge base into their memory aid.

Generally speaking, a good driver will take five to ten laps to understand the layout of a typical modern track. That is, ten to twenty corners. For example, the Monaco Grand Prix track is considered to have nineteen turns.[51]

The first layer is the large-scale geometry of the circuit: the radii of the bends, the lengths of the straights between them, and the overall proportions of each corner. These are still broad strokes. Is the next corner a right or a left? Is it a fast bend or a hairpin? Does it lead to a straight or another corner? Is it uphill or downhill? Is there banking or adverse camber? Already the first lap(s) will inform the driver of this general layout, and also the braking distances required to slow down enough to make it around the corners, and what speed and therefore which gear to take each corner in.

(This requires that the driver must already have a good understanding of the particular car he is driving, and a good feeling for its braking and handling performance. The concept of understanding what the car will do and will not do is dealt with in **Chapter 5** on control.)

Take for example Tabac, Turn 12 at the Monaco track (**Figure 3-1**). This is probably a familiar location to many TV viewers, where the cars pass by yachts on the iconic Monte Carlo waterfront.

A racing driver, of course, sees it somewhat differently from the TV viewer. And not just in terms of their vantage point down at the track. For them, Tabac is 'a superfast left' that you approach at a high speed (F1 cars reach over 230 km/h in the 200 m straight before the entry to Tabac). On approach, the driver will have to drop down a gear or two and be cautious not to be overambitious with the throttle, because you'll push wide and hit the outside wall. But if you are too cautious you end up losing too much speed and that's not good either. Tabac opens up and then leads straight into the swimming pool section, which is a sequence of two chicanes. Back in the day the first chicane was lined with walls. Nowadays drivers can attack it much harder, as the chicane has been opened up a bit. In a modern F1 car it is taken pretty much at full throttle.

Figure 3-1. Turn 12 at Monaco, known as Tabac, involves a high-speed entry into a fast, blind left-hander that leads directly into left-right – right-left chicanes. (Where the track goes around a swimming pool, hence the name 'swimming pool section' for T13 - T15.) Kerbs have to be taken at high speed between steel barriers only inches away on either side. This sequence of corners is possibly one of the most impressive displays of modern F1 cars... and what is needed to drive them.

What makes Tabac superfast is not the speed itself... it's the proximity of the barriers.[52] With no real run off, this is a sequence that has to be nailed perfectly. And if you get Tabac right, then you're going to arrive at huge speed into the swimming pool section.

Once basic understanding of things like this is achieved, and ideally in the least amount of time possible, then the next level is the details of each corner. If we could see into a racing driver's mind, we could peel back the layers of all this track knowledge, like an onion, and peer into the deeper levels beneath the basics.

There would be the subtle undulations (dips, crests, bumps...). The placement of kerbs and stable landmarks like brake markers or marshal posts seen in peripheral vision. Where is there new asphalt? What aggregate did the asphalt layers use? How does this affect the interaction between the tyres and circuit? Where are there bumps that can upset the car? How hard can each kerb be driven over (in this particular car)? Where is the zebra crossing at the apex, and what increase in grip level will it give relative to the bare asphalt? (In the dry – in wet conditions it becomes super slippery.)

Already during relatively few laps, the driver's brain is logging tremendous amounts of this kind of localization and mapping data. Only once all this is understood, for every phase of every corner, for every track, for every car, is the driver able to exploit to the full the performance of any given racecar.[53]

For a typical circuit this data needs to be logged and memorized for a dozen or so bends – the Nürburgring Nordschleife is legendary as it has well over a hundred, many of them interconnected in complicated sequences of dips, crests, banking, blind entries and changing radii: a unique challenge for memory![54] A professional driver will know hundreds of circuits, and from this you may begin to appreciate the breadth and depth of the knowledge the humble racing driver's brain must be carrying inside the helmet.

REFERENCE POINTS

The standard way to describe the knowledge of the track layout is in terms of 'reference points'. These can be any fixed-in-place structures or texture patterns that can be clearly seen from the car. Places where there's a gap in the painted white edge line or a blotch of sand or a building or a marshal post at the side of the track can all become reference points, for example. Or a different colour in the Armco barrier where the fence has been fixed. Or a specific part of a kerb, or a crack or a patch in the asphalt itself... Anything that is fixed in place and visible can be used as a landmark. These give the racer's brain a 'fix' on where you are on the racetrack. They allow the racing driver to place their car and their waypoints, such as braking points, on the track.[55]

The process of learning a circuit builds on these visual reference points. A cognitive map of familiar objects and locations gets stored into long-term memory banks (perhaps somewhere in the temporal lobes of the brain; you'll find out whereabout these are in your head when we get to **Chapter 6**). It is commonly believed that expert racing

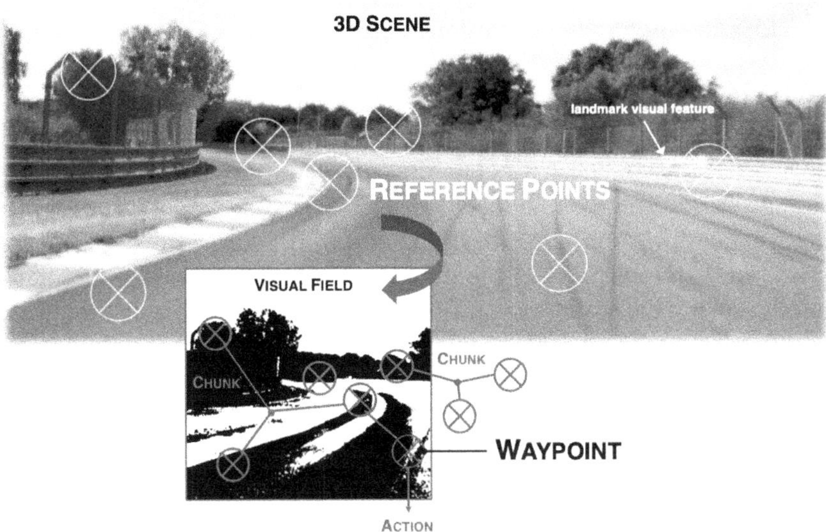

Figure 3-2. Reference points tell racing drivers where exactly on the racetrack they are. They may be chunked in visual awareness into meaningful relational combinations, that are stored into and recalled from long-term memory. (See below for an explanation of the concept of chunking.)

drivers will very quickly learn a very large number of reference points that give them a good sense of their location on the track.

Again, this is not a very conscious process. It's more like when you go to somebody's house, and your brain picks up a large number of 'reference points' – doorways, windows, furniture and so on – so that you very soon have a feel for where you are. And you definitely recognize rooms you have been in… even though you cannot necessarily explicitly say what it is about the physical layout and the objects there that you 'know' when you 'know where you are'. This is not just an analogy – I (OL) do think there is every reason to believe that the very same structures and processes in the brain are behind both cases of spatial memory.

Once the driver has understood the overall layout of the circuit, they can start pushing the car… and the more puzzle-solving aspect of the game begins. What is the best technique for each bend to exploit this car's power and handling characteristics? How should multiple bends be stringed together to maximize speed over the entire lap? What would be the best racing line through a more 'technical' corner where the traditional line is not the optimum? What line through this corner provides the overall best solution?

For example, particularly abrasive surfaces may warrant a more tyre-saving line that reshapes the way the bend is taken, across all the phases. The ambient conditions make a difference, too. How much does the track change with changing temperature? How much does it change from 'green', at the start of the very first practice session for the weekend, to 'rubbered in' at the end of the race? How do the grip levels in different parts of the circuit vary? Is the circuit particularly windy, and what is the current wind direction? How will this affect braking distances, or mid-corner aerodynamic performance?

This amount of knowledge may seem surprising, but it is consistent with the most prevalent finding of the studies that psychologists and cognitive scientists have conducted on the cognitive basis of expert performance in the past 40 years. The exceptional performance of elite performers, even in 'physical skill' domains like sports, is not just a reflection of superior sensorimotor and physical prowess or exceptional innate ability. These may play a part, but in no field of serious competition are they sufficient to take you to the top.

Experts and non-experts have more or less similar general capacities for body movement and higher brain function. What is radically different is that experts go to extraordinary lengths and spend huge amounts of time acquiring knowledge and skills. And it is largely this huge body of domain-specific knowledge that makes consistently superior expert performance possible.[56] This has been found originally in chess – and there are clear and perhaps surprising similarities in chess and racing, to which we turn next.

LIKE HIGH-SPEED CHESS – LITERALLY!

Racers sometimes describe racecraft and strategy as being like a game of high-speed chess. This may actually be much more apposite than they realize. Indeed, the high-speed chess analogy applies even to the apparently much simpler ability to negotiate a track at speed. This ability of the racing driver to select the correct 'moves' and the ability of the chess expert to select correct moves both depend on the same thing: fast pattern-recognition, based on knowledge held in long-term memory. Knowledge that is more detailed and more accurate than most people would probably imagine.

That this should be the case for such a paradigmatic 'thinker's game' as chess is one of the more startling findings in the research on chess expertise. So, let's restate it just to be clear: superior performance *in chess* largely depends on fast pattern recognition and knowledge in long term memory – not powers of reasoning, not 'intelligence', not general 'brain capacity'. It's all about things like recognizing, in a split-second, the promising situations from the sketchy ones. Reasoning is done mainly to check and back up the initial intuition. So, the cognitive basis of performance may not be as different as you might first think, when comparing Kimi Raikkonen with Garry Kasparov.

But maybe this should not be so surprising. After all, the way the human brain is designed is that it will leverage prior experience to support current performance (more on this in **Chapter 7**).[57]

What a human brain needs, in any field of performance, is a tremendously large knowledge base which is, paradoxically, also tremendously quick to access.[58] Such a knowledge base has to be built on hours and hours of experience. This is the same for all fields of expertise – it's just the details of what exactly needs to be stored in the knowledge base that differs.

There is even an analogy that can be made between the moves in chess and the 'moves' you need to make a racecar go through a corner: correct moves made in the right order give you a solution to the puzzle (checkmate, pole position). And this is not just a superficial play on the different meanings of 'moves'. It is an analogy based on deeper similarities in the role that *expert intuition* plays in the selection of those moves (really: the role that long-term memory plays). Learning the skill is a process of *searching* through the space of different moves (actually: the space of different ways of choosing moves).

Of course, there are some differences as well. One difference is that before they commit, chess players may take up to several minutes to check the potential consequences and relevant alternatives of their move. So, there is obviously an additional element of explicit step-by-step reasoning in chess. In race driving, once you enter a bend at speed you can no longer decide at leisure how long you want to spend making up your mind!

Another difference is that the pieces and the moves in chess are defined by explicit verbal rules. The 'moves' in racing are not. Even if some techniques have names and conventional verbal descriptions in the jargon, the verbal descriptions do not completely and exhaustively specify the move, the way, say 'pawn to e4' does.

But actually, in the expert's brain even chess knowledge is not organized around concepts at the elementary level at which the chess *rules* are laid out. (Single pieces and their rules of movement – 'the king may take one step in any direction' etc.). When a person develops expertise and insight, the mind extracts some higher-level, more abstract patterns – which, we shall see, may not be as dissimilar between chess and racing as you might think.

Finally, exactly how you *physically* make your move, whether you handle the pieces with poise, balance and feel or not, makes in chess no difference to whether you win or lose[59]. In racing it is crucial.

But in this chapter, we are still concerned only with the role of knowledge held in long-term memory, and how it supports expert intuition – we will get to poise, balance and the smooth handling of the instruments later. To understand the matter at hand in a bit more detail, we need to look at some of the actual psychology of chess, as discovered by over fifty years of top-level experimental psychology and cognitive modelling; chess being the domain of expertise on which by far the most detailed scientific knowledge of human perceptual-cognitive expertise has been gathered.

SIMILARITIES AND DIFFERENCES BETWEEN RACING AND CHESS

At this point you could reasonably ask: is there an element of fitting expertise in racing into a bit of a procrustean bed by choosing to compare it to chess? After all, if chess is the *only* genuinely well-understood example, what else is there? But bear with us, and we'll try to show that you can glean a lot from chess studies to understand the business of race driving in terms of fundamental memory organization.

To me (OL) this connection has always seemed a very intuitive one, actually. Not that I would have any real skill in chess or in racing, but on theoretical grounds. When I was an undergraduate university student – in theoretical philosophy, no less – I did some way-back-when sim racing stuff (Papyrus IndyCar Racing titles and then Grand Prix Legends). Some of my fellow philosophy undergraduate students, apparently more impressed by my knowledge on the Scottish thinker David Hume than that other Scottish thinker Sir John Young Stewart[60], were bemused at this 'un-academic' choice of genre; at least compared to chess, or, among computer games, what they presumably saw as more appropriate genres, like puzzle games. But I always felt keenly, and tried to explain, with varying degrees of success, that in fact these *were* puzzle games. A crucial part of achieving a better lap time was to study a corner, perhaps download a track map and some notes from the internet, and figure out new ways to try to 'unlock' speed. (There's also the question of skills needed in visual guidance and control – or lack thereof, my excuse for not being very quick despite thinking hard!)

I found this process of analyzing the bends also not that dissimilar to exercises in mathematical logic, part of the philosophy curriculum. You know where you need to go (what you want to prove), where you are (the axioms, and the premises of the proof) and what you need to figure out is how to make them connect in a way that is allowed by the rigorous, unyielding laws of logic – ideally in the most efficient, elegant way. In a bend you know where you need to go (you want to come out the other side), you know where you are now (approaching the bend), and these need to be connected in a way that respects the rigorous, unyielding laws of physics – ideally in the most efficient, elegant way!

In what are now classic studies[61], chess experts were shown to be superior in quickly encoding relational information. Relational information means that what you know about the chess position is expressed in terms of how the pieces are placed relative to one another – not just the coordinates of the individual pieces on the individual squares of the board. A chess position has up to 32 pieces arranged on a board of 8 x 8 squares, yielding an astronomical number of different possible arrangements. (By astronomical we mean, technically, ridiculously large. Never mind what the actual number is. It's too big to comprehend.) This means that any particular arrangement that you ever end up in will not be one that any player has ever seen before, even after centuries of intense study of the game.[62]

What the researchers did was something called a *delayed recall task*. They first showed players chess positions they could study for a few seconds, and then it was whisked away. The players' task was then to place pieces on an empty chess board to recreate the position they had just been shown. The researchers then measured recall performance in terms of how many pieces were placed correctly on the board.

What they found was that the stronger the player, the higher the level of performance in this memory task. (In chess, player strength can be determined quite precisely using a ranking system called Elo numbers). However – and here is the crucial insight of the experiment – this advantage of expertise would *only* hold for real chess positions, from actual games between skilled players. If instead the configuration the players had to recall was a random placement of pieces on a chess board, then the experts performed hardly any better than novices. (It is relevant to stress here that these were not famous games that would have been known to the participants.)

The take-home lesson is this: not all of the astronomically many arrangements are equal. Some arrangements of the pieces are *meaningful* in terms of threat, protection, balance of power, control of the board centre etc. in a way that a random arrangement is not. These properties, the wholistic *meaning* of the relationships between the pieces on the board do not depend on the position of any individual piece. It emerges from how pieces are *related* to one another, and to the empty squares in between. As do the lines of play that the piece's placement affords or prevents. A single piece at any location on an empty board cannot attack, threaten, protect etc. The pieces only do this in the context of all the rest of the pieces.

Here I (OL) think one can now see an analogy to the reference points and waypoints on a track, or in the visual scene as seen by the racing driver. A single reference marker or landmark does not mean anything: it is only the *pattern of relations* between multiple reference points and waypoints, and the empty space between them and the 'lines of play' joining them, the different racing lines, which begin to give them *meaning*.[63]

Now, this analogy should not be pushed too far. When they are preparing for games, chess players probably spend much more time on thinking about chess theory and explicitly trying out different lines of play against a computer and then analyzing the consequences. Nevertheless, in a human-to-human game situation the chess expert does rely – much more than you might think – on just *looking* at the pattern and *seeing* these emergent relationships on the board...and the most promising line of play to make. (Especially high-speed chess is in this way more 'intuitive' than 'intellectual' in character).

So, race driving, or at any rate the navigational-level aspects of it that we are talking about here, can be likened to super-high-speed chess to a degree you might not expect on the basis of surface appearances. The basic design of the human brain is the same in both tasks, however, and whether an individual is developing expertise in chess or in racing – or in cooking, music, writing or gardening – there is a shared fundamental cognitive basis. What is it like?

SEVEN GENERAL 'LAWS' OF EXPERTISE

Psychological research on experts and expert performance has discovered some interesting general truths, which seem to apply to any field of expertise; 'general laws' of expert performance and its acquisition. One technical review of seminal work in the 1970s and early 80s (in experimental psychology and artificial intelligence)[64] summarizes five basic facts which have stood the test of time. These 'laws' seem to characterize the core features of expertise regardless of the particulars of a specific field (**Table 3.1.** overleaf). They reflect more the way the human brain achieves ability in a domain, than the specifics of any single domain.

Let's look at them briefly, in a non-technical way, and see how these general principles apply to motorsport.

1. Continuous development of competence – you never stop learning

First, there is plenty of room in the brain. There is no known capacity limit to humans learning new skills and facts. You can always learn something new, go that little bit further. You can always edge that little bit closer to perfection. Ask a world champion and they will tell you they are still always learning – they never or hardly ever get everything 'perfect'.

One reason this is possible is the unlimited capacity of human long-term memory to store patterns and procedures.

These patterns and procedures are called *chunks*. Chunks can lay dormant in storage, waiting to spring into action when the situation calls upon them. The number of chunks that can be stored in long-term memory has no known upper bound. And because how finely and accurately you are able to chunk the situation is an important determinant of performance, there is no limit to improvement.

You can always learn new tracks and new tricks, no matter how long you race; the learning curve never flattens out completely. There is, though, what economists call 'diminishing marginal gain on investment'. The early part of the learning curve is steeper, the performance improving in leaps. E.g. your laptime may improve whole seconds; you can suddenly 'find half a second' in just one turn, by trying out new things. Then you get to the expert level, and you will be investing a lot of time and effort to gain those last tenths or hundredths. The margins of improvement will diminish.

This is because the new chunks become just very, very subtle variations of a theme, rather than whole new melodies. But within that, you can always do better. The brain will not run out of long-term memory capacity.

2. Domain Specificity – you only know what you have experience of

Expertise is highly domain specific. A great cook is probably an average tennis player, a great tennis player is probably average at chess, a great chess player is most likely to be an average cook. You can only be an expert in a domain in which you have put in the hours and effort towards continuous development of your abilities.

It is not the case that individuals with superior general intelligence[65] will immediately excel at everything they do. Everybody needs to put in the hours. Nor is it the case that if you have gone through the trouble of having developed skill in one domain, say, chess, that you will be able to transfer much of that skill into another domain, say, cooking or tennis.

Race driving is a domain unto itself, and sufficiently far removed from everyday driving that there is relatively little transfer. There is some transfer in the other direction: from race driving and especially visual anticipation and car control skills to dealing with traffic situations. But a racing driver who has never driven in traffic will still need to learn the rules of the road.

3. Meaningful information – you simply 'see' situations as a whole

The match of relational patterns in memory to relational patterns in the situation, chunking, is what determines expert performance.[66] But here the unlimited character of long-term memory is in stark contrast to the severe capacity limits of attention and working memory: how many chunks one can *simultaneously* match and hold active in awareness. This is estimated to be as little as three or four chunks at a time.

TABLE 3.1. GENERAL 'LAWS' OF EXPERTISE
(FOLLOWING GLASER, 1985)

1. Continuous development	You need not ever stop learning: *'There seems to be a continuous development of competence, as experience in a field accumulates'.*
2. Domain specificity	You only know what you have experience of: *'Expertise seems to be very specific'.*
3. Meaningful information	You simply 'see' situations as a whole: *'Experts develop the ability to perceive large meaningful patterns. This pattern recognition occurs so rapidly that they take on the character of the 'intuitions''.*
4. Procedural Knowledge	Expertise is knowing-how, not knowing-that; knowing what to do, not necessarily knowing what to say: *'The knowledge of experts is highly procedural'.*
5. Performance 'not based on thinking'	You simply do what needs to be done: *'These components of expertise enable fast access pattern recognition and representational capability that facilitate problem perception in a way that greatly reduces the role of memory search and general processing'.*

But because the chunks can be big – wholistic relational patterns, entire forests, not individual trees, as it were – the expert is able to encode a lot of information into those 3-4 chunks. (But only in their domain of expertise, which they have developed specific chunks for, not in other types of situations). The novice cannot do this; the infor–mation available is less meaningful, and so the novice is not able to make the same observations and inferences as the expert.

This is the underlying reason for why expertise is so domain specific: you only will have access to stored patterns, chunks, from your domain of experience.

Your memory patterns for encoding chess positions are of little help when you are trying to play tennis. Your chunks for mowing the lawn are of little help when you need to play the guitar.The chunks that enable you to design and execute a Michelin-star meal, given some basic ingredients and utensils, will not help you drive a car.

This itself is directly a consequence of how your brain is designed to work: long-term storage and retrieval of memory patterns is cheap and fast, but encoding new patterns (learning) is expensive and time consuming. The recognition of familiar patterns is efficient analysis of novel patterns is hard work. Habitual responses are automatic, and flow without effort as the situation calls upon them – novel responses require conscious attention, and slow and deliberate, effortful think–ing.

You can probably appreciate why it is the cheap, fast, efficient processes that you need to cultivate to develop a high-speed skill, like race driving.

4. Procedural Knowledge – knowing-how not knowing-that

Knowledge that can be expressed in words, pictures or gestures are what psychologists call declarative knowledge (**Figure 3-3**). Knowledge of how to *do* something, which can (only) be expressed in skilled action, is procedural knowledge. Knowing how to ride a bicycle is a good example of procedural knowledge: you know how to do it, but unless you are a bicycle or motorcycle engineer, we can almost guarantee you would not be able to describe accurately the way a bicycle is kept upright (by steering) and steered (by countersteer–ing).[67]

When you have developed a procedural skill, the information embedded in your brain circuits is overwhelmingly in the implicit procedural form, not in the explicit and declarative form. While many experts can often talk (almost endlessly) about their topic, still most of the knowledge behind performance itself is procedural, not declarative (and very little if any of the declarative knowledge is theoretical knowledge, the kind of knowledge that can be expressed in terms of precisely definable, technical concepts – only an engineer or a scientist needs to develop this particular type of knowledge).

This is the case for any domain of expertise, even those that are commonly perceived to be more 'intellectual', or dependent on acqui–ring 'book knowledge' in addition to procedural knowledge. These fields rest on a solid basis of procedural knowledge as well. The diffe-

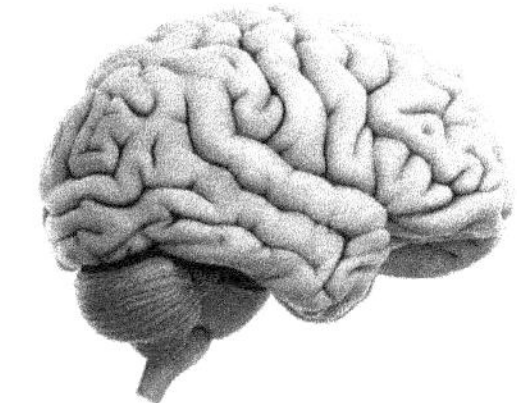

Figure 3-3. From the brain's point of view there is a fundamental difference between procedural and declarative knowledge. They can be roughly (if not completely accurately) characterized as knowing what to *do* vs. knowing what to *say,* respectively.

Procedural knowledge is about performing skilled action sequences. It takes a lot of repetition to acquire and to fine-tune it into a skill. What is encoded in procedural memory cannot often be expressed verbally, only 'expressed through action'. (E.g. how to hit a nail true with a hammer, how to ride a bicycle). In other words, it is *implicit.*

Declarative knowledge is knowing what some object is (e.g. that it is a car), what it is called (in English: 'a car', in German: 'ein Auto'), and what semantic world knowledge is associated with it (e.g. that some cars need fuel). Declarative knowledge is often but not always, *explicit,* i.e. it can be expressed in words.

rences in how much 'book knowledge' there exists in a field – or how much you are expected to learn – makes less of a difference than you might perhaps think. Because *whatever* the domain, the expert must not just know what to say 'about the field', they must know what to do 'in the field'. Mere book knowledge will not get you anywhere, in any domain.

A novelist or a writer should generally read a lot – but just reading will not make you a writer. Mere book knowledge does not even produce books! You have to write.

And scientists will need to study techniques and scientific problems in their field, and how they were solved in the past; but to learn to solve new problems, the as-yet-unsolved problems, you need to do hands-on problem solving. Likewise, chess experts will spend years understanding chess theory and studying openings and endgames and problem situations – but if they want to succeed in competition, they need to play in competition.

The same is true in race driving; even though here the amount of known theory and book knowledge is relatively small. So, perhaps it's more like learning professional cooking – you need to go work in a kitchen and learn to take the heat! Or perhaps it's more like learning to play jazz in clubs, than like learning music theory in a conservatory...

5. Performance is not based on thinking – just 'doing what needs to be done'

The expert will know what to do – or at least have a good idea of what might be a good thing to do – immediately. Just by looking at a chess board an expert player will see a potential threat, a potential line of play, shifts in the balance of dominance. Only then may explicit, thoughtful – slow and deliberate – reasoning be engaged; this is done in order to check whether or not the 'hunch' was right, or if there might be some non-obvious things you have to consider further.

And this kind of search for different 'angles' is very limited – in fact, it is precisely the strength of the expert *to be able to limit* this type of search. (As opposed to, say, strength lying in the ability to search more broadly.)

The classical work on expert chess players has shown that whereas computer chess algorithms can process hundreds of millions of alternatives before making a decision, a human player concentrates on only a handful – but these are usually the *best* ones!

This aspect of human intelligence is in contrast to the classical artificial intelligence approach[68]: there is very little search, or the search is constrained to verify or reject only those lines of play that were immediately suggested by intuition. This is what makes an expert fast and efficient when presented with new, but familiar, situations [69].

It is not necessarily the case that the expert thinks about nothing at all, however. The racing driver who does not need to think about their driving will do well to *reinvest* their freed-up cognitive capacity into thinking about race strategy, car set-up or other task-relevant information (as opposed to, say, dinner plans).

TABLE 3.2. TWO ADDITIONAL 'LAWS' OF EXPERTISE (FOLLOWING SCARDAMALIA & BEREITER, 1993)

6. Reinvestment of cognitive resources	Not just automatization; effort released by making recurring performances routine is put back into the task.
7. Progressive problem solving	Not just efficiency through problem reduction; taking on the task at a higher level of complexity and demand.

This *reinvestment of cognitive resources* freed up by automatization and proceduralization of skill, and the resulting *progressive problem solving* may be at the heart of developing expertise in any domain; seeking optimal performance by learning to tackle or re-interpret the problem at hand on a more complex level.

This is opposed to problem reduction where your goal is satisfactory performance as 'efficiently' as possible, i.e. with minimum effort. Which is why thousands of hours of everyday *experience* does not necessarily give you *expertise*. Experience and expertise are not the same thing!

We may therefore add two more general laws to our list (**Table 3.2.**):

6. Reinvestment of cognitive resources – not just automatization

The development of expertise is not just about the accumulation of routine performance. When performance becomes routine, it becomes automated, and no longer needs as much attentional capacity to per–form. The crucial difference on the road to expertise as opposed to the road to everyday experience is that the attentional capacity freed up by automatization gets reinvested into the task itself, to tackle it at a higher level of understanding or more strict standards of success and failure.

Thus the pursuit of expertise becomes a path of *optimizing* performance (using all one's capacity and all one's resources to perform at the highest level possible), as opposed to *satisficing* performance (using as little capacity as possible to perform at a satisfactory level;

here you may want to quicky refer back to **Table 1.1.**, p. 22).

7. Progressive problem solving – not just 'efficiency' through problem reduction

By re-investing freed-up cognitive resources back to the task and choosing to take it on 'at the next level', the to-be-expert is engaged in something called progressive problem solving. This means that the nature of the problem changes over time, as experience and skills develop.

Even when faced with the same physical situation – e.g. approaching a bend at full throttle on the straight – the novice and the expert are solving different problems. The criteria of success are different, too. (Survival vs. pole position, say.) Purely because of differences in how their minds are perceiving and interpreting that situation.

These processes are at the heart of the development of expertise; indeed, they can be thought of as a kind of definition of expertise itself, as this kind of ever-deepening problem-solving process. As opposed to some particular amount of experience one must have, or some outcome level of performance one needs to achieve.

When viewed as a process, there is no magic threshold of practice hours or ability one needs to cross, to suddenly become declared 'an expert'. What expertise is, in essence, is the expert-like approach to problems in one's domain. The goal posts keep changing, what it takes to achieve them does not.

Over time, a lot of time, this process will accumulate enough knowledge to enable superior performance.[70] This is nicely illustrated by the experience of Andrew Shovlin, Director of Trackside Engineering of the Mercedes Grand Prix team. He has been there to see first-hand how both Michael Schumacher and Lewis Hamilton, tied at the top of the all-time record of seven Formula One world championships, go about their work.

The observation he makes is that at some point the motivation ceases to be less about winning, recognition or other such external factors. It becomes more intrinsic – the value of the activity starts to come from doing the task itself better, as well as possible. From trying to be a more complete driver, year-on-year. To be as good as you possibly can be. *'When you've got that inherent desire … That has just kept pushing the level up, and eventually you just leave people behind. That's what* [Hamilton has] *done'.*[71]

PROGRESSIVE CHUNKING OF REFERENCE POINTS

At a level of underlying cognitive mechanisms these laws, especially laws (3) to (7) directly relate to the psychological mechanisms of chunking. Because chunking is such an important concept to get right, let us spend a few more moments on explaining exactly what a chunk is, and how cognitive scientists (studying chess) originally came up with the concept.

Experts perceive complex relational patterns as 'simply out there', or 'obvious'. But these patterns are subtle, and far from obvious to those who do not have the same powerful pattern-recognition machinery in their heads. The relational pattern of information stored in long-term memory is called a chunk. The term chunk itself actually came from the specifics of the delayed recall for chess positions we described above. How?

In the experiments, they were using a physical chess board with physical pieces. (Rather than a computer screen and a mouse as would be the norm in psychology labs today – this was before the era of personal computers.) What the experimenters noted was that the part–icipant was grabbing a *chunk* of pieces, then placing them on the board in the appropriate squares, then pausing to look at the board, and then grabbing another *chunk* of pieces and so on...

The basis of the chunking theory was the realization that pieces were not being grabbed randomly or based on where they happened to be lying on the table: the pieces belonging to a chunk were selected and placed on the board to create a *meaningful* sub-pattern, before moving to the next chunk. Say, a castled white king surrounded by a rook and a few pawns, a stereotypical defensive array.

The scientists proposed that chess experts memorized the board by recognizing these recurring patterns, chunks. Even if the whole board configuration was not recurring but novel, it was nevertheless built out of recurring sub-patterns which could be chunked.

The more sub-patterns and the bigger sub-pattens you have lying in wait in memory, the more pattern matches there will be for you. This way, a large amount of information can be rapidly encoded, even though, as was already known at that time, attention has a strict capacity limitation of only a few chunks at a time. (This capacity limitation of attention seems to be a part of the cognitive architecture of the brain – something that is *not* modified by training or expertise, but a fairly fixed feature of human cognitive architecture, the current estimate is that it seems to be around three or four.)[72]

Computational cognitive modelling work based on the chunking studies further led cognitive scientists to estimate how big the number of chunks in memory would need to be to support the level of recall seen in the best experimental participants. It turned out this was on the order of tens or hundreds of thousands of stored relational patterns. If you do the numbers – say, twenty reference points for twenty corners for a hundred tracks – you'll find that you're in the same ballpark.

Given plausible assumptions about how fast a player would be able to acquire and store new chunks during learning, it would take years, up to ten years, to reach that level of stored chunks. This is again consistent with the biographical data about time it takes to reach elite levels in chess – or any other complex domain[73] – and gives a principled explanation of why the going is so slow. In chess, as in racing, the old adage holds that there is no substitute for seat time.

According to the chunking theory, what makes an expert is not powers of intellect and deduction – even in a game that on the face of it seems designed to test powers of intellect and deduction. What the expert has in hand over the novice is instead her much, *much* bigger store of relational patterns stored as chunks in long-term memory. Not superior reasoning or superior attentional capacity, at least if attentional capacity is measured in terms of how many chunks you can attend to simultaneously. However, the expert will have superior capacity in terms of how much *stuff* they can fit inside one individual chunk. The expert's chunks can be bigger! They can therefore cram more relational information into the same number of chunks.

And this is what allows the expert chess player to have the expe–rience of seeing *the whole board and everything in it*, in a single glance. Similarly, seeing the whole scene and everything in it is also how a racing driver should train themselves to view the scene:

'It is something you should definitely practice while driving in the street, and also in all other activities of your life. For example, while driving down the highway and using your vision as you normally do, ask your brain for as much information as possible. Ask it to be aware of everything along the side of the roadway. Make note of the ground, the grass and the trees in great detail as you pass it by. But don't just note that they are there. Also note the colors, the type and amount of

leaves on the trees, the condition of the bark, whether the ground is made up of mostly dirt or of rocks etc., and the speed at which they pass by. When doing this, don't look directly at the ground, grass and trees. Look down the road like you normally would, but allow your brain to take in more information (actually, construct more information from the data your eyes are supplying to it). Ask your brain to provide more information. There is a physical limit to how much your eyes can take in. It is practically limitless what your brain can do with that information. What you are doing is practicing becoming more aware of everything around you, using the same amount of visual data supplied by your eyes.' [74]

What I (OL) think Ross Bentley is here eloquently describing is an experience that can be interpreted as developing richer and richer chunks, to combine visual input and long-term spatial memory.

Theories based on chunking [75] are the predominant approach to modelling expertise in cognitive science; have been for the past 40 years. They have not been rigorously applied in the domain of (race) driving, though.

Yet it is well known that highly detailed memory recall of driving events is a memory skill developed by racing drivers. For example, being able to recall how the vehicle responded and performed at the apex of a specific bend, on a specific lap. It is absolutely needed at the professional level for debriefing with race engineers. (Note that this task is not unlike the delayed recall task used to study chess expertise…)

I (OL) believe a chunking-theory based on reference points is what you get if you put together the established views about the importance of reference points in race driving, and the cognitive psychology of pattern recognition, memory, and decision-making seen in *every other* domain of skill and expertise studied.

It is therefore an interesting possibility that the reference points are based on chunking patterns into navigational long-term memory for track locations. The reference points would define a space of interlocking locations that you are moving into (rather than, say, a moving target that magically happens to travel along 'the ideal line'). On this view, the spatial image[76] in the brain is not one of moving targets, but a sense of a stable space – even though this perception of stability is constructed from highly dynamic clues… this is something we'll be returning to.

Instead of pawns, knights, rooks and queens, the chunked elements would be visual landmarks for reference points. Chunks may be de–composed into relations among reference points the way chess chunks decompose into arrangements of individual pieces. Mea-ningful sub-patterns that suggest a specific 'lines of play', specific waypoints.

I find this hypothesis plausible on both face validity (how racers describe track knowledge) and neurological plausibility (how I think wayfinding information is encoded in the brain).[77] And it gives me a buzz every time I think about it.

But what about learning a new track? It's all very well chunking a familiar corner in a split second by matching what is perceived to what is already stored from prior experience… but how does one store that experience in the first place? How does one chunk a corner on first exposure?

One possibility is that a 'library' of stereotypical curve *templates* is stored in long-term memory – analogous to familiar (sub)patterns of chess pieces. If the racing driver draws upon these when presented with a new track, this would allow the highly experienced racing driver to quickly chunk into memory – and reliably recall after only a brief exposure – complex track layouts.

How important is this navigational-level knowledge? In my (AD) experience, the most common differentiator between average drivers and good ones is that the expert can handle corners which are obscured far better – especially with regards to just how quickly they come to finding the optimum line and technique. Corners where a driver, prior to the braking zone, can already see the apex as well as the exit are less difficult to navigate. Time and time again, when working with new drivers I find they will be able to attack corners to a very high level when the track is visible. When it is not, it would appear these other mechanisms that the true expert develops come into play.

This aspect is a bit de-emphasized in many standard driving tech–nique manuals. But, if I am being honest, I think driving technique plays second fiddle to these mapping mechanisms within the brains of top drivers… I see so many top drivers who seem to have their own idiosyncratic quirks with regard to technique that something else must be going on.

To me (OL) this suggests, in terms of cognitive processes, the ability to recall from long-term memory the unobserved geometry, and the creative ability to chunk. Of course, 'emotional' factors like the con–fidence you have in your internal model come to it as well. (But for a cognitive scientist these emotions are just an expression within your awareness of the mechanisms our brain uses to deal with uncertainty, just as vision is an expression in your awareness of mechanisms that deal with optical information.)

Finally, we want to point out that there is also a more intangible aspect of familiarity of a circuit. Drivers can have circuits they just click with, and for no other reason than somehow feeling at home.

Why do some drivers just find a rhythm on some circuits and struggle on others, while another driver experiences the reverse? These are not necessarily facts about the geometry per se, and so this aspect of familiarity may not be directly accounted for by the 'purely cognitive' chunking hypothesis.

It could be the smell of the place or the memories of past glory… but our cognitions and emotions are not in reality separate and indeed in the brain the machinery for place memory is also a part of the emotional system… And it could also be that a particular track just matches better the chunks (prototypical curves in the driver's mental library), and therefore the fine-tuning process sets off on the right foot – whereas other tracks are a poor fit and consequently harder for the driver to make sense of.

All of this builds a complex picture of what it means for a driver to know a circuit. This multi-layered nuance makes racing drivers' track knowledge such a fascinating phenomenon.

As for Senna and his special relationship with Monaco. It's the most iconic combination of man and place in racing history. His six victories on the streets of the principality highlight his absolute dominance of the layout at Monaco. What did it mean for him to really 'know the circuit'? We'll never know for sure.

Chapter Four

The Power of Foresight (Guidance)

A WHILE AGO I (AD) was sat at a restaurant owned by a friend of mine, who happens to be a world-class percussionist. We were chatting away, as we usually do, about this and that; and we got onto the subject of art. In particular, how you go about getting funding to do it. One thing that stuck with me was when he described how he had had to write down the 'meaning' of his performance. He laughed and said, *'the point of it is to express something you can't write in words!'* This resonated with the discussions I'd been having with Otto on how to put into words – which is what you need to do in a book – something that is indescribable. Your implicit knowledge, where you don't really know what you know. The very reason music has such potency over us is because it allows expression of your ideas and your identity in a way that is not limited by vocabulary, by language. For me this is very similar to driving.

We do live in a world dominated by description. We endlessly debate the meaning of films, lyrics, and other art forms. The reality for a lot of artists and performers, however, is that the process of creating and performing is something very hard to put into words. So, when we ask the question *What do I see when I look at a bend on a racetrack?* I can't help but feel a similar response. It's very difficult to actually describe what a driver's relationship with vehicles and a circuit and the flow of it all really is.

To me (OL) this captures a deep tension in the scientific depiction of the mind. We can describe the braking points, the turn-in points, the apexes, the exits, the undulations and elevations – all the navigational elements of a circuit – and where they physically lie on a particular track. We can even explicate the logic behind how they are connected into a 'racing line'. But all that still describes what is *out there* – not necessarily how it all gets connected together *in here.*

We can also take a leaf out of the vehicle dynamics engineer's book, and talk about how the braking, steering, and throttle control create forces and torques that give the racing driver feedback and allow control over the vehicle's speed, direction and load balance (more on this in **Chapter 5**). But how to relate this to 'feel' and sense of balance and the knowledge of the track, to attentional focus and awareness? This is not straightforward.

The middle level of the navigation-guidance-control hierarchy most directly relates to how the racing driver would describe her experience – and brings out its limitations.

For reference points and for the control actions we can, as said, borrow from the vocabulary of navigation, experimental psychology, engineering and physics. A lot of what happens at the navigation and control levels can be described, to a useful approximation, in terms of measurable physical behaviours. Engineers and scientists can laser-scan racetracks, they can simulate control processes and vehicle responses, and many of the principles are understood.

But what happens in between, how it all comes together in the driver's mind – well, for this there is much less exact knowledge to fall back on, at the current state of science. This is actually a bit of a problem for anyone interested in experimental and theoretical investigations into expert performance, and especially for those of us working on expert-level visual guidance.

Because so much of expert knowledge is implicit procedural knowledge, not explicit declarative knowledge, you cannot get the full picture of what is going on simply by asking the person. The person will not always know what she knows. And even if she did, much of it she could not describe because of a lack of common vocabulary. And: what the person *thinks* she knows can be inaccurate!

There is an amusing anecdote on how fragile and error-prone the explicit, verbally expressible knowledge of 'reference points' and landmarks in racing can be (in comparison to the robust job the visual system does in putting this knowledge to use on the track). It is from Stirling Moss, this time driving C-type Jaguars at Le Mans in 1951.

The puzzle he and his co-driver Jack Fairham were facing was where to place the braking point on the fast, sweeping approach to Arnage corner. The road was lined with trees that looked so alike that they were unable to determine clear landmark reference points that they could point to, and name. They scouted the location on foot, and finally Moss painted in white one of the trees, at the location he believed he was getting off the throttle.

Then the final practice session was cancelled because of rain, so the next time they would arrive at the marker would be at racing speeds, in the race itself…

Moss had the first stint and handed the Jaguar over to Fairham. *'When Jack finished his first stint in the race he was convinced that I really was some kind of driving genius because he found that he could only get round Arnage if he braked as hard as he possibly could when he first sighted that tree, never mind drawing level with it!'*

Then Fairham noted the message Moss had left him in the car… *'Arnage white tree much too late. I am braking 50 yards after curve past Mulsanne.'*[78] This of course begs the question: how accurate and reliable was in fact the consensus on *all the other* markers, which the drivers did manage to agree upon? How much of the consensus was perhaps driven by the fact that a prominent 'marker' was present, roughly at the right location, which would make it easy to remember…?

It's just so easy for your brain to come to the conclusion that, for purposes of communication, 'this must be what everybody is using'?[79] In the absence of clear and quantitative feedback there would be little to challenge the formation of such a consensus…

But there is another aspect the example highlights as well: it's not all about how unreliable the explicit, shared knowledge regarding what the driver thinks she is doing can be. Because once in the car and up to speed, it was *dead easy* for Moss to 'check' whether the assigned marker was in fact at the correct place or not. He *knew* of course where he should brake, by his fine judgement of distance, speed, and the braking capabilities of the car. Yet he could only *access* that knowledge when presented with the specific visual inputs of the actual situation – on track, in the car, moving.

This example is therefore also about how the information processing and memory retrieval required for the accurate control of the car can be very dependent on the specific visual impressions and the physical sensations you get, e.g. on the track.[80]

In contrast, a chess expert is able to chunk a chess move whether it is on a physical board, a graphical representation on a computer screen, written in letters and numbers in chess notation, or read out loud as sound – so the chess expertise is less tied to the optics and dynamics of the physical situation than the racing driver's (which is not perhaps that surprising).

The basic principle stands, though, that you cannot drive a corner based only on the way it visually appears. This is especially so in rallying, where you are not allowed to practice and learn the course by heart. Rally drivers critically rely on pace notes giving navigational level knowledge, including knowledge of curve geometries beyond the field of view and hidden dangers such as a rock in a ditch. The rally driver absolutely must drive to the notes, not just 'by eye' – and many crashes have occurred because of failing to take sufficiently into account the navigation-level knowledge.

In track racing where there are no verbal pace notes, drivers also sometimes get 'coaching' tips from engineers, e.g. when they are told that the telemetry shows the teammate braking later for a turn.

But what patterns does the racer's brain perceive 'in the situation'? And how? These guidance-level processes are somewhat elusive. There is nowhere near as much in the way of a cognitive theory of expertise in visual guidance, as there is on expert memory and decision making. (Remember that in chess the visual guidance of one's hand movements, how to actually physically move the pieces, is not considered part of the problem.)

Nor is there an established body of knowledge in engineering that would describe perception and physiology in the mathematical language that control theory and vehicle dynamics use.

This makes the guidance-level mental processes quite paradoxical: they are closest to how the driver visually and dynamically experiences her surroundings – the contents of awareness while on the racetrack – yet they are the most difficult to analyze and put into words afterwards, in verbal recollection or scientific description.

GETTING YOUR EYE IN

In any sport where you need precise high-speed visual guidance – not just motorsport but alpine skiing, whitewater canoeing... – doing practice runs will help you 'get your eye in'. And this is true even for sports where you are not guiding yourself through space, but a projectile that is not you: snooker, darts etc.

Once you get your eye in, you are more confident about being able to accurately control your instrument. You can judge the situation at a glance; and you can more accurately and easily anticipate the consequences of your actions. Your brain is getting ready to get into 'the zone' or a state of flow,[81] where every action fluently follows the other – quickly, without hesitation or a sense of conscious effort.

So, what exactly happens when you get your eye in? Do your eye movements, your visual strategies, change? How do they actually contribute to performance? And how much is it really about getting your *eye* in, as opposed to getting your brain's visual and motor systems in tune?

Visual strategies figure prominently in race driver coaching. Textbooks present analyses and describe exercises meant to teach how to look at and how to see the racetrack in the correct manner. Perhaps the three most common guidelines for visual guidance are those given in **Table 4.1.** Let's consider them in turn.

1. Look far enough ahead – really far ahead

The first one is a widely accepted idea: you need to look far enough ahead, and you need to develop a habit of looking farther than you might automatically. For example:

'It takes practice to feel comfortable looking farther ahead than you do now, so begin practicing it on the street. You will be amazed at how much it will help – and at how far ahead the winners are looking'. [82]

This is common not only to driving racecars and racing motorcycles, but also driving instruction and safe motorcycling strategies on the road.[83] And it has been shown experimentally by using eye-tracking equipment that novice drivers tend to look at the road closer to the vehicle than experienced drivers.

The idea in each case is that the further you are looking, the more time you are giving your brain to assess the situation and react before you get there.[84]

How far is far enough, however is not quite clear. There could be a 'golden mean', as you might wind up looking too far and miss potholes or other critical scene elements closer up.

TABLE 4.1. THREE 'BASIC RULES' OF A RACING DRIVER'S VISUAL STRATEGY

1. Look far enough ahead – really far ahead
2. Look where you want to go – do not look where you don't want to go (that's where you'll have your crash)
3. Use peripheral vision – 'take in everything' in 'widescreen awareness'

It is also an interesting question whether it makes a difference whether the aim of the drill should be conceived as *physically looking* far ahead, or whether it should be conceived as developing a *mental awareness* of your physical surroundings... and expecting that the eyes automatically follow. The difference is that in the first case you are consciously interfering with gaze control, which is normally totally unconscious, and expecting the appropriate perceptions and situational awareness to follow. In the second, you are changing your attention and situational awareness and expecting the eyes to follow.

Thus, visual guidance training can actually be conceived in two ways: 'Look (far) and you will automatically see', or 'See (far) and you will automatically look'. The latter might seem counterintuitive– how can you see if you don't first look? But remember that most of what you see, anyway, is all the time in peripheral vision; not at the point of gaze you are 'looking at'. Also, the way your brain is wired is that wherever you focus or shift your attention, this automatically becomes a potential gaze shift target for the (oculo)motor systems.[85] Which brings us to:

2. Look where you want to go – do not look where you don't want to go; that's where you'll have your crash

The second idea, really, a set of interconnected ideas, is to 'look where you are going'; or, more precisely, 'look where you want to go'. And to avoid looking where you shouldn't be going... because 'wherever you *do* look that's where you *will* go!' These adages reflect the tight coupling in eye-hand coordination (or more precisely gaze direction and movement direction).

Again, an active visual strategy is recommended here: an unexpected obstacle on the road tends to capture attention, and gaze, and lead to a kind of tunnel vision where you just focus on the threat – and drive straight into it! So, the recommended remedy is to take over the situation by volitional control, and deliberately move gaze away from the obstacle (expecting steering to follow). 'Look and you will automatically steer'.

Or maybe it's again attend-and you will automatically look/see/steer?[86] We are here talking about movement of the eyes, of which we are not conscious, and which happen at a split-second time frame over which our intuitions of what we think we are doing do not always match up well with what we are actually doing. And in racing we are dealing with events that take place at sub-second scale, and under considerable physical and mental stress. All this makes retrospective memory of the order of events to be taken with a pinch of salt.

Although we may *feel* like our 'conscious self' should be able to freely and *instantaneously* take over whatever behaviour we want – control of the eyes, or of the wheel, or of attention – and although it may *appear* that we can recollect the sequence of events with infinite precision, this is actually an illusion. In reality, the brain patches up a kind of 'best-guess narrative of what just happened', based on sensory and motor signals that arrive at various delays, and any fragments of memory that match them.[87]

It has been shown, in laboratory conditions, that estimates of simultaneity (that is, judging whether two events happened at the same time) and temporal ordering (which of two events came first) are to a surprising extent a post-hoc (re)interpretation which the brain puts together from all the data available at the time of recall. Human perception and memory do not work like a video recorder.[88]

So, at this level of detail it is difficult to know on the basis of phenomenology alone whether it is really gaze leading steering or steering leading gaze (or if the question itself is even meaningful to pose).

3. Use peripheral vision; 'take in everything' in 'widescreen awareness'

The third idea says that you need to use peripheral vision, to 'see every everything'; to see the whole scene flowing by without looking direct-

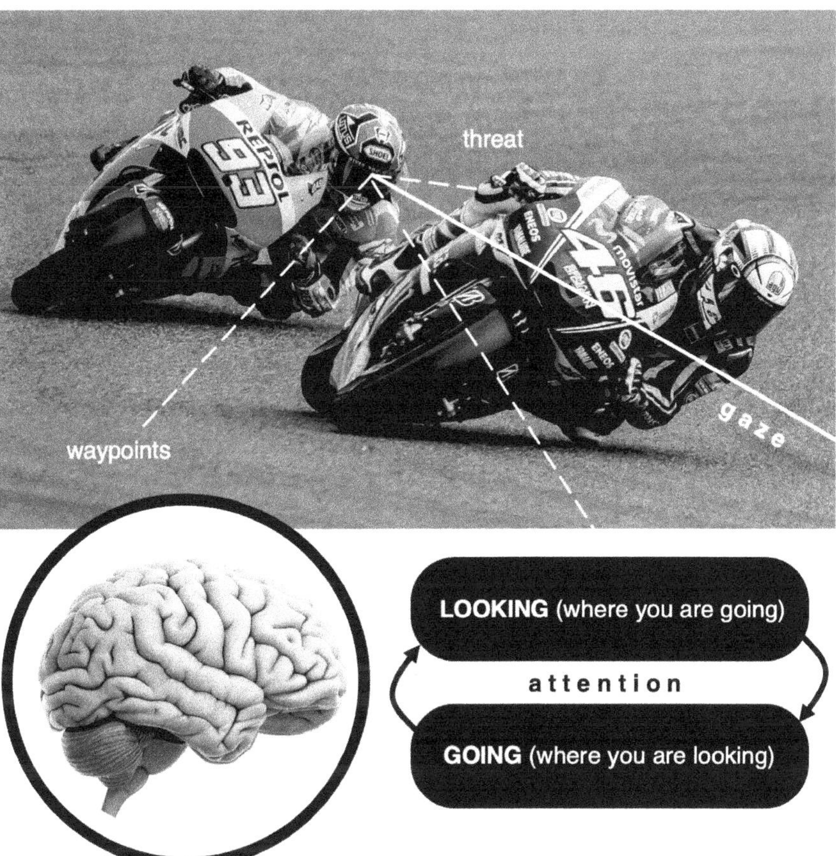

Figure 4-1. There is a natural and universal tendency to look where we are going (while tracking objects or locations in the visual periphery). This is an automatic, habitual pattern. Another equally ingrained tendency: to go where we are looking. Looking and going are said to be coupled, in both directions: you look where you go but also go where you look. These fundamental visual strategies are fine if you're driving/riding into a bend and looking ahead into the curve... but if there is an obstacle on the road it can get you in trouble. The threat will capture your attention and your gaze. (Another fundamental: your gaze tends to go where your attention is.) You become fixated on the threat so that – because you go where you look – you ride straight into it! To avoid this undesirable outcome, motorcyclists, both on the road and the race track, are instructed not just to look where they are going, but to look where they are *supposed* to be going – because 'wherever you look, that's where the bike will go...'

ly at anything other than the general direction where you want to go. Racing drivers are instructed to develop skills in this sort of 'widescreen' visual awareness:

'Do you remember the first time you drove very fast, or skied down a mountain? Your field of vision, or awareness, was probably very small – like looking through a scope. But the more you drove quickly or skied, the more your vision expanded, and the more you noticed around you. … When I first drove an Indy car, my field of awareness was narrowed by the speed at which everything was happening (just as it was when I first drove a Formula Ford, and then a Formula Atlantic car). But as I became more accustomed to the speed, the more my field of vision and awareness expanded once again. Experience in a fast car – at a high speed – will help you become acclimatized to this speed, and increase your 'field of awareness'.'[89]

The idea here is to make your brain – your conscious awareness or your attention – register peripheral visual detail, but *without* overt[90] movements of the eyes. This makes the way the racing driver should 'scan' the racetrack quite different from the way expert drivers in traffic (e.g. driving school instructors) conceive of scanning; the latter emphasize visual search, with overt eye movements, and explain to students they should be constantly *moving their gaze around* to focus on different parts of the scene – pedestrians, traffic signs, red lights, side roads…

The 'widescreen' idea, on the other hand, suggests the absence of scanning. Or does it? Remember, introspection of one's own phenomenology is an unreliable guide to what the brain (or the eyes) are actually doing.

One research paper[91] reports an interesting example of this in another domain: an artist exclaiming how her way of looking at her work is quite different from the everyday, how she is trying to take in the whole painting at a single glance – but when they actually measured her eye movement behaviour with an eye tracker, it displayed the usual pattern of flitting from one scene part to another, several times a second (described below).

You see, our brain is always reconstructing a unified conscious perception from the snapshots of different locations, arriving at different times; yet we are only aware of a single, enduring, flowing scene. Here, however, measurements of racing drivers' eye movements do back up the phenomenology. Instead of overt scanning, racing drivers do appear to exhibit a kind of 'thousand-yard stare' that is different from normal road driving (**Figure 4-2** and **Figure 4-3**, overleaf). There is comparatively little work published[92] on expert racing drivers' eye movements, but even if you look at onboard cameras showing racing driver's eyes (especially clear in simulators, or night racing), you'll actually see how they're most of the time staring at the distance with hardly any to-and-fro (and very few blinks).

So, insofar as they are simultaneously trying to 'take in everything', this must be done using covert visual attention. (This means moving your attention without moving your eyes; this cannot be seen with an eye tracker, and currently there are no precise methods to directly measure covert attention shifts).

What are the covert attention shifts used for? Why are they a good thing? For various interesting reasons that have to do with how your brain processes visual information, the racing driver cannot or may not wish to move their eyes very often.

Keith Code of the California Superbike School well describes this problem:

'[W]hen your eyes move, they ... flit from one object to another like a butterfly. ... A rider's problem is he wants to see the track in front of him flowing as a whole scene, to maintain a steady flow of concentration, but his eyes don't work that way. If he stares at one reference point too long, he'll experience a form of tunnel vision. But because of the way his eyes work, he has to look at some specific thing!'[93]

This contains many good insights into visual function. Our eyes normally do indeed work by 'flitting from one object to another like a butterfly'. Oculomotorists, scientists who specialize in the movements of the eye, call these flits *saccades*. (From the French word meaning a jerking movement, as in jerking the reins of a horse.) Between saccades the eye remains static on a target. In most tasks these *fixations* last for about 0.1 – 0.3 seconds at a time. This goes on almost all the time, every waking hour of your life![94] The normal default scanning pattern in almost all everyday tasks, including driving in traffic, is to make these rapid eye movements that that bring new objects of interest to central vision for detailed assessment.[95]

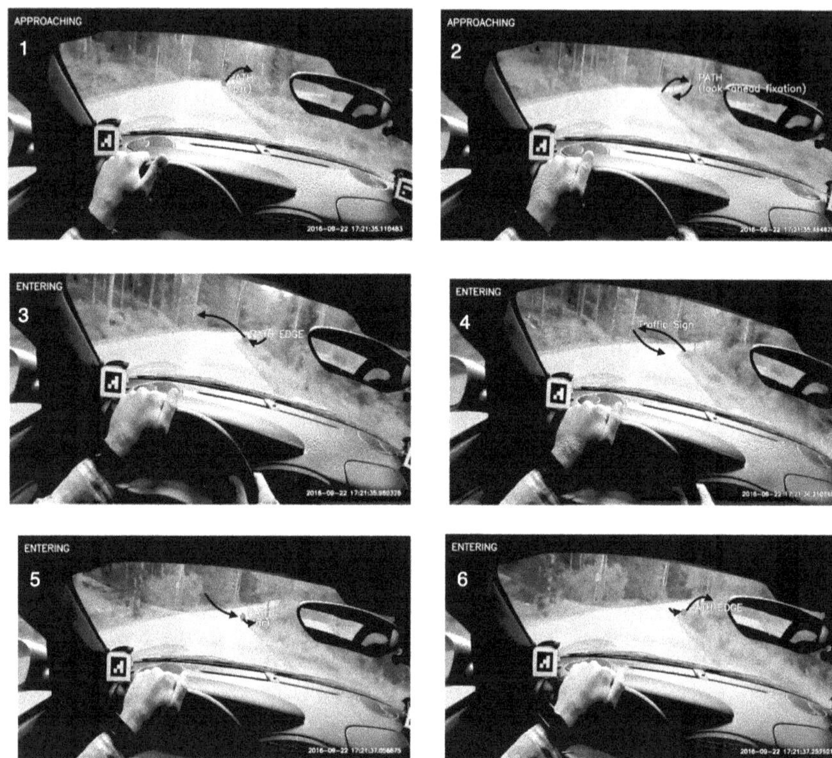

Figure 4-2. Experienced drivers in everyday traffic scan the visual scene with frequent rapid eye movements; this image sequence represents about 2 seconds of scanning. It is adapted from the eye-tracker head-cam movie in Lappi et al. (2017), showing a professional driving school instructor's gaze behaviour.. Note how within this short time the gaze already makes six jumps, saccades, across the scene, scanning relevant objects and locations one at a time. Some fixations are guiding fixations, looking where you aer going but for scanning around, when you are looking where you are *not* going, the coupling between looking and steering needs to be temporarily broken.

This type of scanning strategy may not be optimal for a racing driver, though – for basic physiological reasons that have to do with the way the eye and brain encode visual information (as explained in the main text). Indeed, their sports-specific skill sets include drills for developing peripheral vision and a 'widescreen' awareness without moving the eyes (see discussion in Lappi, 2018).

Figure 4-3. Heat map showing the tight concentration of gaze in the driver's visual field for a professional racing driver on a test track from Lima et al. (2020). Even if (or indeed because) most of the track is seen in peripheral vision, the direction of gaze can be highly focused in a very small region (straight ahead on the straight, around the apex in turns).

The coordination pattern of fixating at the apex is also very clear in everyday driving, but there it is interspersed with frequent scanning of the visual periphery and the road further ahead. How exactly speed and expertise affect a driver's visual strategy is still largely an open research question.

The heavy reliance on peripheral vision in many sports is chal-lenging for researchers trying to understand them, because an eye tracker will only tell you where the eyes are pointing... which is by no means the whole story of where attentional focus is!

Eye tracking measurements have shown experienced drivers on the public road have a broader scanning pattern than novices, implying moving attention around more broadly.[96] Racing drivers may not scan overtly, with their eyes, but this does not necessarily imply tunnel vision if the usual tight coupling between attention shift and eye movement is severed in their visual strategy.

Why might visual guidance and the peripheral vision drills for racing drivers be so different from the natural pattern of everyday contexts? One reason may be the different environment: the traffic environment is way more unpredictable than a racetrack; a child might appear from behind a roadside billboard at any time, the door of a parked vehicle might open, the leading vehicle might change lanes, with or without indicating etc.

Another obvious reason is the speed at which things happen. How does this relate to the fundamental neurophysiology of the visual system? The movement of the eyes itself dramatically affects the incoming visual inputs. Every time we shift our gaze, our brain actually shuts off the input from the eyes for a brief moment. After the 'cut' the next snapshot scene is analyzed.

The same happens during blinks. Yet you don't see the world suddenly going dark during a blink. This is because your brain *commands* the blink, and so it can predict and 'factor out' its effects on the incoming signal… an unexpected lights-out of similar duration (a couple of tenths of a second) would certainly be dramatic!

In both cases the visual signal is shut off for a tenth of a second or two, up to several times a second – but we do not notice it.

Because during saccades and blinks the brain effectively shuts down the transmission of information up the optic nerves, you cannot see your own saccades (or blinks). Look at yourself in a mirror and try it: stand in front of the mirror and then repeatedly shift gaze e.g. between looking at your left eye, then your right eye, then your left eye… you'll be surprised.

If you get somebody to stand in front of the mirror next to you (or record your eye movements on a smartphone), you will find that you can easily see *their* saccades and blinks, so it's not that these events are too quick for the eye to register. It's just your own brain shutting down reception of visual input, for up to a couple of tenths of a second!

You can also observe your saccadic movements (and blinks) very clearly if you very gently feel your eyeball through the eyelid with the tip of your finger. We do not normally notice these movements because what we *visually experience* is the representation our brain patches up afterwards from all these fragmented snapshots, and from a lot of material retrieved from memory.

So, you can see why the way we experience the world is not the most reliable source of information about how the visual system really works. We experience a continuous vision of a stable 3D scene, even though what our brain receives from the eyes are intermittent 2D snapshots, registered during fixations and coming in 2-3 times a second. That means that when we are actively scanning with our eyes, we are typically 'blind' for 10% of the time!

This is a clue to the 'widescreen' strategy: at speed it may not be a good idea to go blind for a few tenths up to three times every second... So, this may be one physiological reason why the brain does not do a lot of saccades (or blinking) on the track.

But even if the eyes are not scanning, attention may still scan around the peripheral visual scene, because attention can visit regions in the visual field covertly, without always triggering overt eye movement.

Another benefit for the racer's brain may be that while it takes a couple of tenths of a second at least for the brain to program each saccade (limiting the rate to 2-4 per second) covert attention can shift around much faster, perhaps up to ten shifts per second, according to some estimates. So, 'widescreen' may involve (covert) scanning after all.

TOP-DOWN AND BOTTOM-UP INFORMATION IN VISUAL GUIDANCE

How the driver or rider gets to 'see the track in front of him flowing as a whole scene' has to do not just with the signals coming in from the eyes (interrupted by saccades and blinks or not), but also very much to do with the memory information already residing in the brain.

Psychologists and neuroscientists call the signals from the sense organs *bottom-up* information, and the signals from brain-internal memory, motor and other processes *top-down* information (**Figure 4-4**, overleaf). Perception is a process of combining top-down information with bottom-up information. It is not simply registering senso-

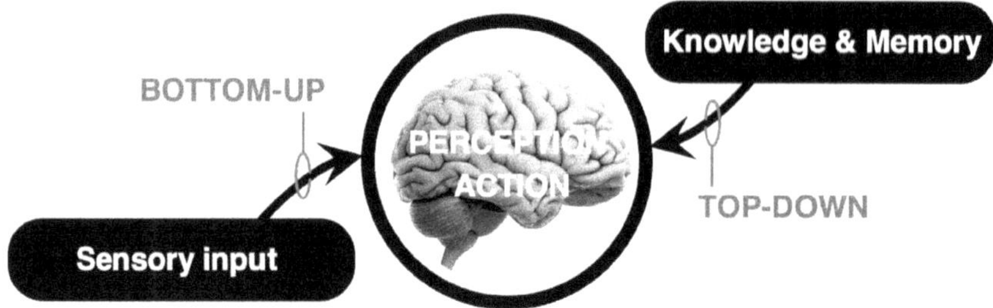

Figure 4-4. Information flowing in from the sensory organs is called bottom-up; information residing in the brain is called top-down. Both are combined in visual perception and the guidance of action.

ry input, or building a picture of what is 'out there' from bottom-up signals only.

For the racer's brain, the top-down information includes reference points in (navigational) long-term memory. Mr. Code goes on to say:

'At speed, the further you look, the slower you believe you're going; the closer you look, the faster you believe you're going. Look far enough ahead to avoid accelerating the scene, but not so far that you lose your feel for where you are on the track … Having enough [reference points] *has the effect of opening up the track, making it appear larger'.*[97]

He also describes detailed practical exercises that give a good idea of the kind of visual routines he has in mind (the same drill is described in other driving manuals as well).[98]

The key to the drills is to deliberately keep your gaze fixed on one location, but not become too fixated on it. In other words, *gaze* is fixed but attention roams around. So, the aim is to develop a habit of 'seeing', but not (automatically) 'looking'! To break the automatic, habitual coupling that where your attention goes, your eyes follow. This takes some effort and executive attention.

This makes sense for many reasons. One being avoiding target fixation from staring at a single reference point that then becomes your waypoint.

This 'widescreen' attention also makes possible what is described as the (optically impossible!) act of seeing around corners. Here is how Ross Bentley describes it:

'One of the most difficult visual challenges you face, especially on street circuits, is seeing around corners. Often, on road circuits – and, with the cement walls lining the track, almost always on street circuits – your view through the turns is restricted. What you need to do is see around the corner. ... It is as if you are building a mental picture to fill in the holes in the visual picture.'[99]

What is happening here is that the navigational (top-down) knowledge supports the guidance level processes which 'piece together' a continuous, flowing visual experience, and make possible a smooth flow of control-level actions.

What is behind the corner – the exit of a blind bend – lies beyond the current field of view. So, the information about it has to come from the brain, from memory, rather than from the eyes. It is associations this navigational level long-term memory makes with incoming visual preview that enables Guidance processes to chunk the scene into a meaningful pattern.[100]

'You must work from the end back to the beginning of the turn to establish your product [outcome]. Decide in advance, before you go into the turn, where you are going next. You must be able to 'see' the product [desired outcome] *of the turn in your mind as you enter it. This enables you to keep the pieces and parts of the turn working toward the product. This overview allows you to figure out each step necessary to arrive at the product or destination.'*[101]

Unlike in closed circuit racing, rally drivers are not allowed to practice the stages long enough to commit them to memory in detail. The driver and co-driver reconnaissance the course and prepare a set of pace notes that contain information about unseen road geometry.

For example, the rough curve geometry (left or right, whether the direction is the same or opposite as the preceding bend, if the curve is particularly long etc.), speed (slow, fast, really fast), changes in elevation (over the crest), the length of the following straight, whether there are invisible hazards like rocks that mean you cannot cut all the way to the inside, and so on. The information is similar to the pace notes of Moss and Jenkinson, **Chapter 1**, only more detailed.

The co-driver will read pace notes aloud at speed, at a rhythm that allows the driver to take in the verbally given information while at the same time controlling the car and guiding it along.

This allows higher performance because you simply cannot drive at a very high speed by 'the way the bend looks as you approach it', and expect to come out the other side unscathed.

This makes rally driving particularly challenging from the point of view of integrating navigational level information with car control, which happens at the level of visual guidance. Because some of the navigation knowledge is now not only retrieved from *memory* – it has to be recovered from *verbally* encoded messages as well.

These messages are just retrieval cues, though. They are memory aides for recalling navigational level information – not exact instructions. The notes are not detailed enough to specify exactly how to drive. They are not as detailed as a musical score or chess notation, for example; it is unlikely that you could drive an unfamiliar rally stage at speed 'prima vista' from pace notes alone and come out the other end!

To achieve speed, it is not of course enough to recognize a bend or to be able to determine the curve geometry – you have to be able to actually drive as well. This means that the prototypes, or whatever the mental items in the curve library might be, need to be not only matched to sensory *input*, but to the control *output* as well. The prototypes must be part of procedural knowledge, and waypoints must be associated with the correct motor programs and muscle commands…

Chapter Five

Power Would Be Nothing without It (Control)

IN THE PREVIOUS CHAPTERS we have been concentrating on knowledge, memory and visual perception. This can give a bit of a 'dis-embodied' view of the racing driver's performances. As if it were a nice indoor job... minus the blisters, bruises and physical pain... minus the hanging on for dear life, holding your breath and hoping your neck holds up (if you're not quite race fit).

The chess analogies we've used bring home some aspects of the task which many people might not realize are there; so it is important. But they do not convey some of the brutality of the sport. If you muck up a chess move, you won't get broken bones, or a concussion (or worse). So, there's an emotional layer there because when it's *for real,* your brain just knows this. (This of course has a neurological basis.)

Since we are here approaching the racing driver's skill 'top-down' – from navigation-level, through guidance, and are only now arriving at control – the skills involved might also have started to seem a bit... theoretical. As if the perfect driver had a perfect facsimile model of the track in their head, and based on this they would hatch a perfect plan, which is then guaranteed to produce the perfect line through a perfectly known corner. Then, it would be just a matter of perfectly executing the pre-programmed motions in an exact sequence – without error, noise, or uncertainty.

This sort of *machine-like* approach to control is not, however, how the human brain, as a biological system, goes about doing things.

When coaching racing drivers, I (AD) notice that with drivers who don't start until they are in their mid-20s and older, they can become very metronomic with braking *points*, steering *points* etc. This is possibly from reading about these in books. Whatever the cause, every lap they'll brake at the same point, turn in at the same point and so on. Whereas more experienced drivers tend to be a bit more... dynamic.

It's hard to explain. But driving a racing car is rarely a case of turning the steering wheel and getting *exactly* where you wanted to go, because you are modulating the brakes as well, trying to load the tyres in a particular manner, which will provide the best grip when you want it. A car at the limit is moving about quite a lot underneath you. So, unless your car is set up really, really perfectly, you are making many corrections.

Rather than literally a turn-in *point* on a track, you may have a small initial turn-in – this will shift the load to a particular corner of the car – then sort of wait until everything is right, and only then put more lock into the steering. If you want to go fast, you have to have the car at the proper dynamic state of (in)stability at all times.

You can hit an apex point geometrically perfectly, but if the car is not also loaded correctly and, more than that, travelling at the right speed and pointing in the right direction, you might find yourself running wide or going too slow or getting power understeer, or whatever. Those would be errors; in fact typical 'textbook errors'.[102]

ERROR AND VARIABILITY

Do the best drivers consistently outperform the others because they know what to do, and then metronomically execute this knowledge, every time? Actually, no:

'A mistake many drivers make, especially inexperienced ones, is trying to drive the perfect line, to the detriment of their speed. Ironically, finding the last ounce or two of speed often comes from not worrying about driving the absolute perfect line. It comes from improvising on that supposedly perfect line and allowing the car to go where it needs or wants to go. ... Learn what happens when you do something different.'[103]

From a car-control perspective there is inherent variability in human motor control. What the great Russian physiologist Nikolai Bernstein called 'repetition without repetition'. It is an interesting phenomenon that I (OL) would say has not received the kind of recognition in the experimental study of driving (and other expert skills) it should.

Not all *variability* should be conceived of as *error*, you see. Humans are not good at machine-like repetition of a preprogrammed routine. It just isn't how our brain is designed to acquire and execute a skill…

If you try to drive with too fixed preconceived ideas in a racing car, you will probably not be as fast as you could be. One reason may be of course that the slow, verbal, part of the brain interferes: what racers refer to as 'overthinking' it. But there may be a deeper and more interesting reason as well.

It is very easy and awfully tempting for a scientist or an engineer to begin thinking of all variability as 'human error' or 'noise'.[104] But this is taking the metronomic, machine-like execution of pre-existing motor programs as the *ideal* case to aspire to. The definition of 'perfect' control.

But when I look at human performance, metronomic execution does not seem like the true mark of skill expertise. One example would be that when you're going really fast, you don't turn in and put on the power at exactly the same point, at exactly the same rate.

This has not to my knowledge been investigated in the driver performance research literature, but a few examples from other domains that have experimentally studied will be enlightening.

Consider how a human being controls the movements of the body, and the movements of a tool that has become an extension of the body. (From the brain's viewpoint you can think of the car as this kind of tool.) The way the control works is not quite what you might at first assume. The classical work here was done by Bernstein on blacksmiths hammering iron.

From the brain's point of view the problem of control can be stated like this: it needs to control movements of the joints and muscles. This is what motor commands directly influence. But indirectly, it needs to control what is happening at the actual points of contact between the hand and the hammer, and ultimately at the point where the hammer meets the anvil.

Here it is useful to make a distinction between contact forces and control forces. The hand exerts *contact forces* on the instrument, the instrument exerts *control forces* on the environment, shaping the metal. This is analogous to a race driver controlling a car, the vehicle being the instrument. The contact forces a racing driver exerts on the car are when he pushes the pedals or twists the steering wheel. The control forces are when a slipping tyre pushes on the ground at the contact patch, moving the car.

So, although the brain can only send direct motor commands to the muscles to change the contact forces, it ultimately wants to guide the control forces the hammer/car exerts on the piece of metal/track.

Where and when and with what angle and how much speed and force the hammer lands on the metal or the rubber meets the road: this is the ultimate outcome that matters. The brain's predicament is that it cannot control the tip of the hammer/the tyre contact patch directly, it only controls the muscles and joints of the arm and the body…

So, you might think that to *accurately* control how the hammer meets the iron, the brain would have to *super-accurately* control the grip forces at the hand, and *super-duper-accurately* control torques from the muscles at the various joints… The deeper you go from the hammer in towards the brain, the less wiggle room for variability there would be. Any small perturbation (variability, error, noise) in the early phases of the movement would accumulate, and produce an even bigger error in the outcome. Right?

To test this idea, the scientists attached small lights on the body and the hammer, and took time-lapse photographs that made the trajectory visible, transforming a dynamic event taking place over time into a static pattern on the photographic plate, which could be measured, analyzed and more easily interpreted. (This was in the early twentieth century, they did not have the benefit of modern computerized motion capture or vehicle telemetry).

To their surprise, they found that the behaviour of the joints and the trajectory taken by the hammer were *more* variable than the end result. The outcome was invariant despite variability in the actions that produce that outcome. In fact, thanks to the variability in the actions, a change in one joint angle was compensated by a coordinated change in the others. Wow.

The same thing was found in another classic study, this time in triple jumpers. This study was conducted in the early 1980s. Apparently the common belief among jumpers and coaches at that time was that a good jump would be preprogrammed during preview – when taking the set – and then executed as planned.

This was possibly influenced by the 1970s motor program theories, which assumed that 'perfect' control in high-speed performance would be about executing pre-stored motor routines (because everything is changing too fast to register sensory feedback).

This would mean that once the athlete set off, the program would run its course so that the jumper would 'hit her marks' perfectly on each step, resulting ultimately in the final take-off heel strike falling as accurately as possible on the board.

But when performance was videoed and analyzed in detail, it turned out that the real picture is quite different. Once again, the smallest variability was in the final heel strike location, and the preceding steps leading up to the take-off step ('release') actively compensated for variability in the placement of footfalls up to that point.

This means that the parameters of movement were still being adjusted, on the fly, even during an overlearned automatic routine sequence. It was not a wholly preprogrammed, ballistic, action.

(This is, incidentally, once again a case of where athletes and coaches do not actually do what they think or say they are doing – even when they are doing it successfully. Due to prevailing theoretical ideas it's very easy to have a factually inaccurate explicit theory of one's own performance).

Can you see how all this is analogous to the racing driver 'hitting the apex', but not through metronomic movements of the hands and feet? And how 'getting the bend right' is not necessarily about driving your own wheel tracks and hitting the same apex exactly, every time? And how this makes the racing line different from the fixed connect-the-dots metaphor from **Chapter 2**? (It's as if the location of the dots magically shifted, as you are drawing, to make a more beautiful picture!)

So, if the variability should not be thought of as error, what is it 'for', then? To me (OL) it suggests a rather deep property of *resilience*, creativity even, that is inherent in human behaviour. A motor plan that is *just* a simple linear 'ideal line' at the navigation/guidance levels, with precisely fixed action points, would be... *fragile.* Any error will propagate to later phases and accumulate over time.

Industrial robots (like the ones putting together cars on an assembly line) are designed based on the original idea that in order to reach an invariant, controlled outcome you should plan the movements precisely, and avoid variability at the source. Here variability is treated as error.

In the racing domain this would mean that ideal control would be like the perfect driver having a perfect facsimile model of the track in their head, and based on this they would hatch a perfect plan guaranteed to produce the perfect line through a perfectly known corner.

But an industrial robot's motor plans are fragile. Humans' and other organisms' motor plans are robust. If human motor control was based on fragile motor plans, we'd be extinct. This applies particularly to racing drivers!

Ultimately, in the real world *nothing* is ever perfectly known because the world is always bigger and more complex than your mind…

A machine is good at performing exactly the same operations over and over again. To perform simple repetitive tasks, we construct hugely complex elaborate machines, and computer programs (which are really a kind of machine).

But machines are bad at adapting to variation or surprises. Think of a robot on a conveyor belt putting together chassis components of a car: the chassis arrives, always one at a time, in the same orientation, with the exact seams to be welded (to this particular make and model) preprogrammed into the system.

There's no need, and no room, for improvisation. But if there are changes, the system needs to be reprogrammed – by a human, the real source of the flexibility. The intrinsic design itself is fragile to environmental variability.

We humans, by contrast, are poor at performing the exact same operation again and again. We get bored. So, we improvise. We try to find better, more robust, efficient or just easier and less energetically taxing (or more fun) ways to carry out the task. But when something new comes along, we can adapt.

'FEEL'

So, what is happening in the racer's brain at the control level? To build intuitions on that, let us turn our attention to the notion of 'feel'. This is really quite fundamental because you cannot have proper control of a racing car without good feel. Sensing what the car is doing at any moment is inextricably linked to influencing it in the next. You cannot have control without sensory information; and 'feel' just refers to how that information is subjectively experienced.

To get a feel for what 'feel' consists of it will be instructive to look at a simpler problem first, instead of a detailed analysis of the really, really complex problem of feeling the balance of a racing car.

We want you to consider instead the (somewhat) analogous task balancing a pen on your finger. This will allow us to build up towards the intuitions needed to discuss 'feel' and control, because you can see more clearly the essential problem when it is stripped from all the complexities in what goes into balancing a racing car's complex non-linear dynamics on the edge.

Like all instruction related to feel and control, this works best if you actually physically do it. So, please find a pen and balance it on your finger. (Physically do it not just 'in your mind' – *don't worry* if people think you're being a bit weird!)

There. Did you do it?

What probably happened is you adjusted the pen with one hand, moving it first perhaps in bigger steps, then finally in tiny nudges this way and the other, and at the same time you changed the orientation of the supporting hand by slightly rotating your wrist and forearm.

Your gaze was probably intently fixed on the pen, maybe to observe if it starts to rotate, and maybe you even controlled your breathing in step with the task. (Take a deep breath – hold – blow out – relax – now put the pen on the finger…)

All of this coordination was about your brain subtly trying to find the point of balance. This was done on 'feel', but much more than just feeling the pen on the tip of the finger through the sense of touch, right? It's not just feeling the contact force on the pen: vision, sense of your arms and body in space, stabilizing your core muscles all play a part in achieving good control. (This should sound familiar if you are a racing driver.)

Physically, what you needed to do was to line up the point of support – the point through which the control/contact forces between the pen and the finger run – with the centre of mass. The point of support needs to be on the line that the force of gravity between the centre of mass of the pen and the centre of mass of the Earth runs through.

In other words, the support force from your finger (lift) must act on the same line as the force of gravity (weight) so there is no torque on the pen. Otherwise, the torque will cause the pen to rotate, i.e. it will experience rotational acceleration.

(A shorter pen with less polar moment of inertia will roll more easily and prove harder to balance, a longer pen has more rotational inertia and will be initially easier to find its point of balance – but once it gets going it is harder to get it back. Again, does this sound familiar?)

Unless you are a physicist or an engineer, and additionally a massive geek, this analytic and explicit way of framing the problem is probably not *at all* what you were thinking about, though.

To get feedback that your skill-learning system can actually work on, you intuitively tried out different things and checked out what happens. You kept going at it until you found a technique of balancing the pen that works. This complex, dynamic, interactive process we experience as 'feeling for the point of balance'.

Note that when you are balancing the pen on your finger, what enters your awareness is a more elusive thing than just the surface texture and pressure, i.e. the touch of the pen's hard surface and weight on your skin – the elementary sensations of the contact forces.

We experience in a more wholistic way where in space the centre of mass, or point of balance, of the pen is; the thing we are trying to control.

We don't experience just how our finger is curled up in space, or the skin pressure at the contact point. We experience apparently 'directly' when we have reached the state of balance. The invariant end goal of the activity.

This directness is only apparent, though. For all this to work, many parts of your brain have to be incredibly busy in the background; compiling, sorting and comparing information. Sensory systems generate masses of data.

The pressure of the pen on the skin is registered by touch. The position and orientation of the arm and hand, the tension in the muscles and tendons, is reported by proprioception (the sense of body in space), as is the direction of gravity (your sense of vertical).

Your gaze was focused on the pen, feeding in visual signals about the relative positioning and movements of the pen and the hand, but also about which way is 'up'. (But try it with your eyes closed – you will see that, unlike racing, it is not a task that is absolutely critically dependent on vision!)

The 'feel' we experience, apparently so directly, is in reality a sophisticated combination of all these sources of information, and more, being put together in the background.

Let's take another example: try to explain *in words* how you sit down into a chair. It is hard. Much harder than just sitting down. Most of us can't do it, really. (Try it.)

Procedurally we 'know', or our brains know, that the support forces from the ground define force vectors which again need to pass through the center of mass. This is why you lean your upper body forward.[105]

You may at this point want to try to sit down very slowly and observe yourself doing it. (This time people will definitely think you are being weird, though.) Slowing things down spares cognitive capacity for observation, and you'll be able to see 'what your body knows' about these things.

If you have studied physics, and only if you have studied physics, you will know all this explicitly as well.[106]

CONTROL THEORY

Trying to use language to explain what a racing car is doing, or what the racing driver is trying to do, is equally hard. In fact, it's the same problem. So, racing drivers have their own jargon: *'The car is loose'*, or *'I get snap oversteer'* – but even so explaining in words what constitutes 'feel' in a race car can be a frustrating exercise in trying to describe the indescribable.

There are some established expressions everyone in motorsport will be familiar with: *'It's like the car is an extension of your body'*, *'It's a question of balance'*, or, a little less poetically, *'You need to feel it through your arse'*. Like with many turns of phrase that become cli–chés, there is an element of truth in them. (You cannot take all of it too literally – your 'arse' in this case resides mostly in your head, for example… there's probably a really bad joke in here somewhere…)

But when the racing driver says they feel 'one' with the car, there may be *more* truth to it than you might guess. Because the brain has sophisticated means of keeping a record of what objects in the world actually belong to the body, are part of 'you', and what objects don't and are part of the 'outside world'… and this record-keeping system is surprisingly flexible (we'll return to this in **Chapter 8**).[107]

To supplement the limited expressive power of spoken language you will often see drivers also gesticulating dramatically. There is some strange hand-waving produced in the pits when talking about going fast! There's a whole array of gestures and stylized movements that come with learning this 'body language'. It conveys *what* you are doing and *how* the vehicle responded. Sometimes these come complete with sound effects, for added emotional effect and precision on the timing. *When* or at what *rate* the action and the response are taking place.

Engineers need to be able to interpret all this 'racer speak' and metaphors and hand waving and the noises, and translate them to 'engineerese'. The race car will not understand metaphors, it only understands numbers. Engineers also use concepts developed for 'engineerese' to communicate to one another and the computers they use to run CAD and CFD and FEM models, so that they can design and develop the car. Computers don't 'get' hand-waving, they only 'get' mathematics.

Therefore, the basis of 'engineerese' is *control theory.*[108] It is a mathematical language to describe control in physical terms. This is a useful thing to have in hand also when you try to understand *the brain function underlying car control* in physical terms. This language gives a more analytical grip on the situation.

Control theory does not say how the racer's brain specifically, or the human brain in general for that matter, might control a system (be that a racing car or any other system). Control theory is a very general framework for discussing *any* systems in terms of their dynamics and feedback loops.

Familiar to automotive, aerospace and race engineers, the theory applies to any system equipped with the means to regulate a dynamical system's behaviour. Control theory is used not only in the design of automobiles, but also aircraft, power plants, mobile robots etc. It is used in ergonomics to model 'the human controller'.

In biology, control theory is used to analyze hormonal and neuronal feedback control. It is also used in computational neuroscience and cognitive modelling to develop models of the brain.

It yields a common vocabulary for all of these, and principles to unite phenomena in a principled way. In doing so it pins down some concepts that will be very useful to us, for understanding more precisely the concept of feedback that is so central to 'feel' and learning from experience. It gives you mental reference points to get a better fix on your ideas.

DRIVING FROM A CONTROL THEORETICAL POINT OF VIEW

Control theory itself is based on a theory called dynamical systems, which is about the coordination patterns and interactions that arise when systems are influencing one another over a period of time. When the output of one system at any time is the input to the other and vice versa, the systems are said to be *coupled*, and causal influence is circular, rather than a one-way street.[109]

In our case, the systems of interest are the racing car, the ground, the atmosphere, and the body of the racing driver – all applying forces on one another, over time.

The physical foundation of locomotion is the controlled application of Newton's second law. You push the ground backwards, the ground pushes you forward. Action and reaction. Control forces between coupled systems. Because the Earth is so much more massive than you or your car, your effect on the Earth is ignorable, but you and the car will hurtle forwards.

In running, you do the pushing with the sole of your foot. In a race car the tyre does the pushing, but the basic principle is exactly the same. To accelerate to a higher speed you push back some more, to slow down you push forward, and to steer you push sideways.

Friction between the ground and the soles of our feet is providing the control force that allows you to steer a path. How hard you can accelerate and turn your body/car in any situation depends on the friction available, the grip and the traction.

(How big control forces the foot/tyre can generate at the contact patch are proportional to the coefficient of friction between the materials, the size of the contact patch, and the vertical load on the sole of the foot/ on the tyre.)

When you walk or run, forces proportional but opposite to the control forces you exert are experienced directly upon the limbs and body. When you push on the ground, you can feel the pressure at the sole of your foot, and the tension in your muscles.

In vehicle assisted locomotion, the control forces are not exerted or experienced by your body directly. Instead, control is mediated by contact forces at the contact points between the body and the vehicle:

where your hands grip the steering wheel, where your feet push the pedals, and where your backside 'feels the g-forces'. These contact forces are not equal and opposite to the control forces. (Just like the force at the tip of the hammer is not the same as the forces on the palm of the hand.)

The contact forces and control forces are said to be transformed into one another by the dynamic properties of the vehicle. You brake, and the reaction forces at the contact patch are transformed into contact forces transmitted to the body at the pedals, seat, steering wheel and seatbelt contact points. You extend your right foot (just a slight increase in contact force at your big toe), and a tremendous control force is generated at the rear wheel contact patch.

(And something primitive in your brain goes: '*Yeah, baby – that's what I'm talking about!!!*' – but on this control theory and physics are mercifully silent.)

Or consider steering: in a car you do not steer by adding a lateral component to the push of the foot, at the contact patch, the way you would steer your body in running. You do not use your feet to push a pedal with a sideways force. Instead, you steer by exerting with your arms a contact force that creates a torque on the steering wheel (we are for the moment ignoring the fact that really you steer with the throttle as well).

Likewise, you feel the 'bite' of the tyres through fluctuations in the steering wheel's self-aligning torque, which is being transmitted by the contact points. A driver can feel almost as if they were touching the ground themselves.

(This is probably neurologically exactly the same as using a stick, as a blind person would, to feel an object – you feel the texture and the pressure *at the tip of the stick,* not in your fingers! Try it. We'll get back to you on this.)

The physical separation of contact forces from control forces not only means that the control force magnitude of the 'extended body' – the vehicle-body system – can be substantially greater than what can be generated by the human body. More than that, the *pattern* of forces, and the *pattern* of outcomes of actions are also different in terms of which action at which limb produces which type of reaction from the vehicle/body system.

These new patterns of relations have to be learned whenever you want to learn to control a new vehicle (car, bike, plane, boat, skateboard…). But remember: what the human brain is particularly good at is *pattern recognition* – finding out 'meaning' in relational patterns it is presented with. Also, most of the guidance and navigation skills needed will probably be the same, so all that kind of knowledge can be recycled from one locomotor context to the other.

This is pretty amazing when you think of it. You can put a four-year-old in a kart and they 'get' it. They get a 'feel' for the thing, and therefore they can control it. The new mappings between the forces that your body generates and experiences (contact forces) and the forces that actually change your speed and direction (control forces) are internalized without thinking.

This implies some kind of adaptive modification on the part of the central nervous system. Where and how this adaptive modification takes place in brain tissue, we'll get to in a moment. The end result, however, is that their brain becomes able to interpret as meaningful feedback all the physical sensations the kart is generating. Feedback is the currency the brain trades on for control and learning.

FEEDBACK & FEEDFORWARD CONTROL

What precisely makes a sensory signal 'feedback'? And how exactly does feedback relate to control? These concepts derive from control theory, and it will be instructive to take a quick look at the conceptual basis of the technical notion of feedback, because the concept of feedback will be so central to the picture of the brain we'll start building up to, in the next chapter.

(Although, as we shall see, brain science uses it with a twist; you will find that the same word, 'feedback', is used in a slightly different meaning in neuroscience, which can be confusing. But what can you do? To understand control in the racer's brain you have to understand both control and the brain… there's no way around it.)

In the vernacular, we routinely use the word 'feedback' for a number of related but rather broad and fuzzy meanings. 'I got great feedback from the audience' (probably meaning they liked what I said) or 'can you give me feedback on this book manuscript?' (meaning positive or negative evaluation).

The technical engineering notion of feedback is related, but more specific. It means that *values of a controlled variable are fed back to the controlling system to be compared against a reference value.* (For

TABLE 5.1. DIFFERENT MEANINGS OF THE WORD 'FEEDBACK'

Everyday, vernacular use	*'The coach gave me great feedback on my technique.'* *'Some drivers can give really detailed feedback on setup changes to the engineers.'*
Control theory definition	Observed value of the controlled variable fed back to the controller. E.g. measurement of vehicle yaw rotation fed back to an electronic stability controller. Note that this concept can be applied both to the human driver and automatic stability control/steering systems.
Neuroanatomical definition	Signals in recurrent connections from higher brain areas back to lower areas (e.g. from association-areas dealing with more abstract object features and scene layout back to primary visual cortex dealing with edges and contours (see the next chapter for an explanation of 'higher'). Note: anatomically, lower areas are said to feed signals forward to the higher areas – even though from a control point of view those signals carry feedback information!

the neuroanatomical definition **Table 5.1.** gives a sneak peek – what it means will be explained in the next chapter.)

So, what does feeding back a controlled variable value really mean? And how does it relate to feel and car control? To see this, you need to look at the car-and-driver system the way a control engineer would, as a connected network of dynamic systems.

The driver's brain is a system that transforms feedback (and other information) into motor commands. The driver's body is a system that transforms motor commands into contact forces. The car is a system that transforms contact forces into control forces. The environment then generates feedback (as well as other signals that the driver's sensory systems can register).

The analysis of control systems[110] separates a controller, this could be the driver, and the controlled system (a.k.a. plant), this could be the car.[111] The controller system takes some input, this could be a pre-programmed set of instructions specifying what actions to take, sensor feedback readings, or some combination of these… it takes whatever input it gets and transforms it into a control signal for the plant. For example, the input could be a driver's decision to slow down, and the control signal would be the motor commands sent to the foot to increase the pressure, a force, on the brake pedal. The output would be vehicle speed. How much deceleration for each newton of force on the brake pedal? That will depend on the design of the braking system (plus many other variables we are ignoring here).

If there were multiple plants, the controller could send out multiple signals that the different plants could 'understand'. Each plant then takes its control signal, and transforms this into an output, according to its internal dynamics.

If the output of the plant, deceleration, has no effect on the controller at all, then the control is open loop control.[112] But if the driver is able to sense the change in speed (he should!) then the output is said to be fed back to the controller. The fed back output becomes feedback, part of the input at the next time instant. (How soon the change can be registered depends on the sensitivity of the driver.) The feedback closes the loop, and so this is called closed loop control (**Figure 5-1**, next spread).

The very simplest form of closed loop control is pure feedback control (**Figure 5-2**, next spread). This is when the feedback is compared against a desired reference value for the controlled output variable, and the difference, called error, is used to produce a proportional response; the bigger the error, the bigger the control signal.[113]

Let's look at the braking-in-a-straight-line example from the point of view of these concepts. Approaching a turn, the driver's brain might have some reference speed that experience tells him the car needs to be slowed down to in order to make the corner.

In response to pressure on the brake pedal, the car – the plant – outputs speed and acceleration. (The acceleration is proportional to the brake pressure, but the speed depends on the previous speed as well as the acceleration, so it is a different transformation.) The controller – i.e. the driver – can sense both. They could be visually sensed, from optic flow or reading the speedometer, through hearing the engine sound and wind noise, or felt from vibration frequencies and contact forces on the body. Actually, it's a combination of all these and more, stay tuned.[114]

Open loop control

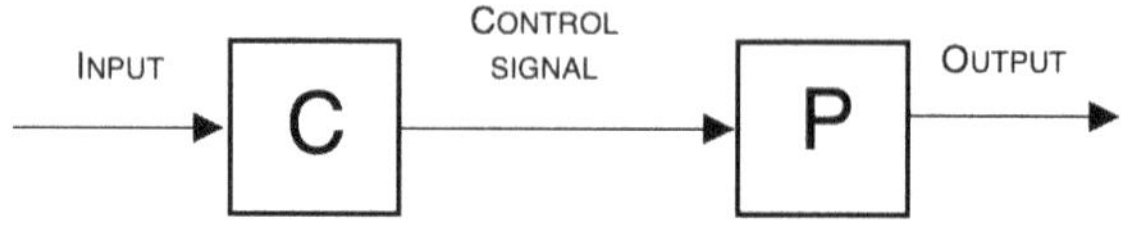

Closed loop control

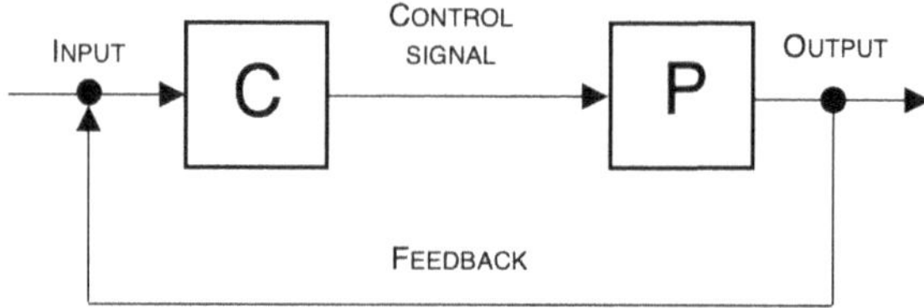

Figure 5-1. Open loop control and closed loop control, and the concept of feedback in the context of control theory. The controller (C) transforms inputs into control signals. The plant (P) transforms control signals into outputs. If the change in controlled variable is observable, the new value is 'fed back' to the controller and can be used for closed-loop control.

Example: when the driver presses the brake pedal (control signal), the car (plant) will respond by slowing down (output). If braking was a ballistic motor program, it would run its course independent of output (open loop control). For the racing driver, however, the sensation of speed (feedback) makes a difference to how one gets off the brakes.

Let's say that whilst approaching the corner the current speed is higher than the desired turn-in speed, the reference. That means error exists (in the technical sense of the term – it does not mean the driver is doing something wrong). Therefore, the controller specifies a motor command to the plant to increase the pressure on the brake pedal, thus slowing the car down.

The next observation of speed should be a lower value, so the driver can ease the pressure on the brake pedal a bit.

This is an example of negative feedback control. It's called negative

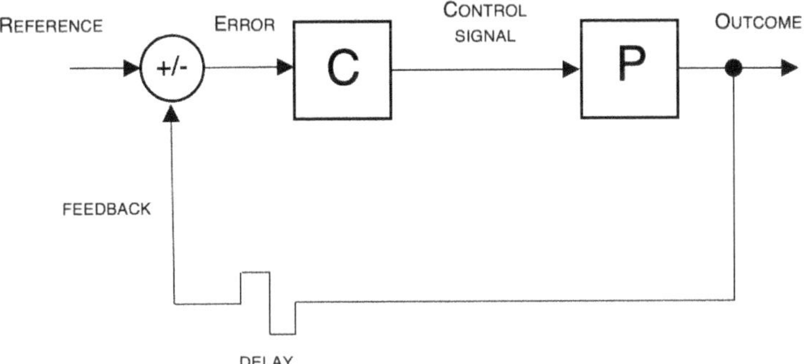

Figure 5-2. Simple model-free feedback control. The controller compares the feedback signal to a desired reference value and transforms difference (called the error) into control signal.

This kind of control is always *reactive*, because the error can only be computed – and therefore the change in control signal generated – after the feedback has arrived. Because there is always a physical delay, which for the brain can be quite long compared to the speed at which situations develop, this can lead to overshooting, overcompensating and oscillatory behaviour: you brake, slow down too much, speed up again, have to brake again… typical novice behaviour in a race car…

Reducing feedback delays and making the controller more sensitive are possible remedies open to the engineer (they do come with their own trade-offs), but for the human brain physiological constraints set some hard limits. The brain has some clever ways to deal with them, however.

feedback because the control signal is opposite to the effects of any perturbation, and will make the controlled variable tend back towards a reference value.

(Positive feedback would be a self-amplifying runaway process. It's what you get when you put a microphone next to a loudspeaker – any random perturbation or sound gets amplified by the positive feedback loop. Another example of positive feedback is if you try sitting down

Figure 5-3. A thermostat in an example of a feedback controller. It measures the difference between desired coolant temperature (the set point) and observed coolant temperature (feedback). When there is a negative difference (error) the engine is still cold and the thermostat will remain shut, blocking the flow of coolant, thus causing engine temperature to rise. As the engine warms up, the difference reduces until it crosses zero difference, at which the thermostat opens, and the radiator begins to cool the engine. Feedback about engine temperature is used to drive the coolant flow.

Note that while we are talking in informational terms, as the thermostat 'observing' the value of temperature and 'comparing' it to a 'goal' set point, in a car radiator this operation is carried out by thermal expansion and contraction when the thermostat's metal components are heated/cooled. But the way neurons 'register' and 'compute' information is similarly based just on their biophysical properties, not some mysterious powers only brain stuff would have.

in a chair without bending over and shifting your centre of gravity: there is a torque arm from the support point on the ground to the center of mass, which has a component horizontal to the ground – not just perpendicular where the centre of gravity and the support point vertically align. This creates a rotational acceleration… which takes the center of mass and the point of support even further apart, horizontally, increasing the torque… again a positive feedback loop that quickly escalates.)

Negative feedback works exactly the way a thermostat in your house's or your car's radiators do. Thermostats are an example of negative feedback control engineering.

Another example of negative feedback control could be the homeostatic maintenance of core body temperature at a reference value of 37 degrees Celsius. Cooling down causes the hypothalamus

– a deep part of your brain involved in body state regulation – to send out commands that, through activity of the musculoskeletal and autonomic nervous system, warm up the body (e.g. shivering). Conversely, warming up makes it send commands that cool the body down (e.g. sweating).

But as you might have noticed with house radiators, a thermostat will click on and off, on and off, on and off. This is because your thermostat measures water temperature in one part of the circuit, while controlling the heater element of the boiler at another part. This means there is a *delay*.

If the thermostat senses that the temperature is less than the reference value you have set (on whatever scale is marked on the control knob), it turns on the boiler. But by the time the heat has flowed out of the boiler and got registered at the thermostat end, the boiler end may already be a little bit too hot. Yet the thermostat keeps everything on, because it does not yet sense what is happening at the boiler end.

Similarly, the boiler end will get cooler than the desired value before the thermostat turns it on again, because it is only responding to delayed feedback. So, the temperature overshoots the reference value going both ways, leading to an oscillating behaviour.[115]

Delays in the system are a fundamental problem for pure feedback control, because this type of control can *only respond* to such delayed feedback. Fast and accurate sensors and quick-responding plants can go some way towards solving this. But a lot of the design of control systems really is to develop cleverer ways to control the plant; to make the system stable – not oscillate too much with overshoots and overcompensation – but also responsive.

In the case of the racing driver, if pure feedback control was all the racer's brain was capable of, this could lead to an oscillating modulation of speed. Slowing down too much, having to accelerate again, then slowing down again… Obviously a skilled racer's brain is a better controller than that.

Part of the skill comes from accuracy in speed perception, and from fast responses – but there are inherent feedback delay limits, based on the maximum speed at which signals can travel to and from the brain and how fast the chemical transmission of messages between brain cells can be. To overcome these problems the *(Continued on p. 118)*

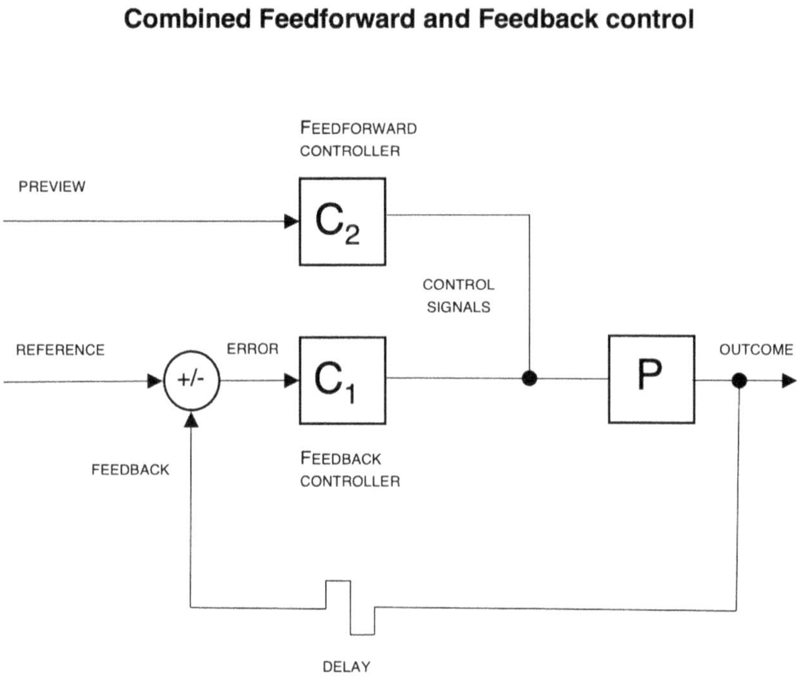

Combined Feedforward and Feedback control

FEEDFORWARD CONTROLLER
PREVIEW
C_2
CONTROL SIGNALS
REFERENCE
ERROR
+/-
C_1
P
OUTCOME
FEEDBACK
FEEDBACK CONTROLLER
DELAY

Figure 5-4. Combined feedback control (C_1) and feedforward control (C_2). Feedforward control allows the controller to 'respond' already before error has had a chance to accumulate, in an anticipatory manner. But only for predictable outcomes, and if preview information is available. Feedback control is there to compensate for unanticipated (unpredictable) perturbations.

Because humans are able to perform skilled acts at such a high speed that there is not enough time for feedback to reach the brain, given physiological delays, some form of feedforward control is necessary.

It used to be thought that a lot of motor skill learning for fast, sequential movement would be compiling ballistic 'motor programs' that run in open loop control – effectively reading out the correct motor commands from a motor routine stored in memory.

Another way would be to preview prospective signals from the environment, and respond to those in prospective control. Actually, the brain's feedforward controllers are a bit more sophisticated than either of these (as will be explained in **Chapters 7 & 8**).

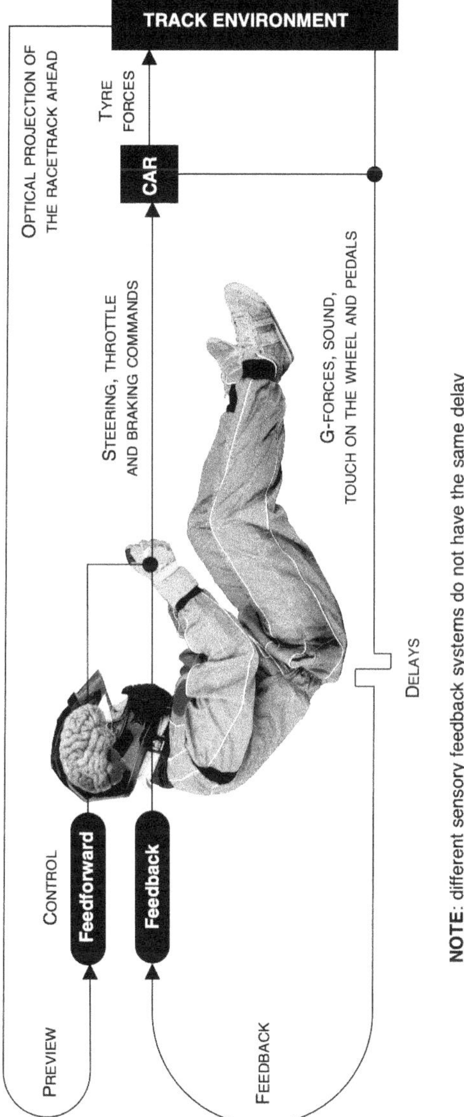

Figure 5-5. The racing driver as a combined feedback-feedforward controller. For example, steering is often modelled as a 'two-level' control architecture with optical preview of the track ahead driving feedforward control, complemented by feedback compensation for bumps, wind gusts and/or limitations (uncertainty, error) in the feedforward controller's internal model.

(Cont'd from p. 115) brain, and also modern control engineers, often use more sophisticated control principles than pure feedback control.

In fact, even the basic homeostatic regulation of body temperature in organisms is not just thermostat-like feedback control. As soon as you enter a sauna, for example, your body starts to mobilize cooling processes. Or, closer to home for the racer: already when you open the door of a hot closed-cockpit GT car for a stint, your body is preparing its thermoregulation to cope. The 'response' occurs even before your core body temperature has gone up.

This kind of anticipatory control is, from the point of view of the controlled variable (core temperature) *feedforward*. By registering some *preview* signal (the open door of the GT car and the heat on your face), the controller is able to send out a control signal that pre-empts perturbations in core body temperature. Even at this totally unconscious level of basic bodily function your brain is a machine built for anticipation.

A feedforward controller of coolant temperature in a car engine, as opposed to a simple thermostat, could preview variables like engine rpm, throttle position or the ambient temperature (these are variables that have predictive value with regard to future engine temperatures). It would then adjust coolant flow based on those variables, rather than the output temperature itself. Those are variables the controller has no direct control over; it only has access to coolant flow (control signal) and thereby temperature (output), so from a strict definition point of view these other observable variables are not feedback.

This sort of control (prospective control) enables adjustment of temperature before a change in temperature, but in response to other variables which are not controlled by the controller.[116] Note that the control system's internal decisions are still just blind reactions, what engineers and neuroscientists call model-free control; it is not genuinely 'predictive'. Yet, because it is purely responding to preview *variables,* the responses can nevertheless 'anticipate' changes in temperature! Still, such a controller is not really *predicting* the changes in temperature. It is responding reactively to preview (e.g. engine RPM).

There are other ways, more intelligent ways, in which the brain can use feedback and the other signals: mechanisms that go beyond simply comparing feedback to a fixed reference value or responding to preview cues. These more sophisticated and interesting ways will in fact blur the neat picture of 'feedback is for closed-loop compensation and feedforward is for anticipation' – but they are important because they give the brain huge possibilities.

Both feedback and feedforward control above are called model-free control, because the controller is not storing internal models of the environment into memory. In *model-based predictive control*, on the other hand, the controller does use internal models for (feedforward) control; it will then use feedback primarily to teach it better models, as opposed to driving control signals directly. Model-based control is, arguably, what brains have evolved for.[117] (Or at least they are a realistic way to describe the sophisticated ways in which the brain is 'wired for prediction'.)

But to relate all these ideas to brain function, we need first to take a closer look at the actual design of the brain. And for that we need to begin a new Part of the book.

'IT OUGHT TO BE GENERALLY KNOWN THAT THE SOURCE OF OUR PLEASURE, MERRIMENT, LAUGHTER, AND AMUSEMENT, AS OF OUR GRIEF, PAIN, ANXIETY, AND TEARS, IS NONE OTHER THAN THE BRAIN.

IT IS SPECIALLY THE ORGAN WHICH ENABLES US TO THINK, SEE, AND HEAR, AND TO DISTINGUISH THE UGLY AND THE BEAUTIFUL, THE BAD AND THE GOOD, PLEASANT AND UNPLEASANT. ...

IT IS THE BRAIN TOO WHICH IS THE SEAT OF MADNESS AND DELIRIUM, OF THE FEARS AND FRIGHTS WHICH ASSAIL US, OFTEN BY NIGHT, BUT SOMETIMES EVEN BY DAY;

IT IS THERE WHERE LIES THE CAUSE OF INSOMNIA AND SLEEP-WALKING, OF THOUGHTS THAT WILL NOT COME, FORGOTTEN DUTIES, AND ECCENTRICITIES.'

- HIPPOCRATES (CA. 460–377 BC)[118]

PART III

THE BRAIN

Chapter Six

A First Look Inside

INSIDE THE HEAD it is dark, warm, wet and quiet. Incredibly faint electrical signals are whizzing around, forming intricate patterns of activity. It is when those signals get shaped by the environment through sensory input – in the ways information is actively sampled by the brain through motor commands and then woven together with prior brain activity – that we have access to the world. And rather miraculously – in ways that are not yet understood – from those signals our knowledge and understanding and, yes, even conscious experience emerge.

Move your point of observation by just a few centimeters, and outside the racing driver's head it's all noise, heat, visual stimuli; the world whizzing by at breakneck speed. (Racing drivers do not wear a HANS Head And Neck Support System for fun.) Yet, in the end, it is only those faint electrical signals that give the brain any access to the outer world of speed, sound and fury.

Let us return to our racing driver from **Chapter 1.** We left him at the start-finish line, 'throttle pinned, engine pulling cleanly and screaming to the redline'… Rewind, if you will, the picture a little bit. See him in the final corner, coming onto the straight. And let us also slow time right down; this will allow us to see more clearly what is going on.[119] And this time we will be armed with all the take-home lessons from **Chapters 2** to **5**. We have been looking at the job description of the racer's brain[120]. It's now time to see how the brain gets the job done, and the actual machinery that does it.

If you looked inside the body of the driver, you would see sensory nerves delivering feedback from the sense organs into the central nervous system. You would see motor nerves carrying motor commands out to the muscles. It's in between these that the magic happens – the navigation, guidance and control processes (**Figure 6-1**)...

We'll start with the business end of the brain: motor control. This involves the transformation of different sorts of sensory signals into muscle coordination. We'll ask first: how does the brain bring together different muscles to perform a single action? We'll then proceed to the way the brain sorts out different sensory signals: how does it interpret them as useful perceptual feedback?

Let's begin with a very concrete question: how does the driver push the throttle pedal? It goes something like this: muscles in the calf extend the ankle because they are being sent motor commands from the spinal cord. These motor commands are electrical signals travelling first from the brain to the spinal cord and then down along nerves into the leg. There the nerves make synaptic connections with muscle fibers and excite the extensor muscles at the back of the calf. And these muscles contract. Simultaneously flexor muscles on the frontside of the calf are sent inhibitory commands to relax. This is the clever bit. The end result: a torque around the ankle joint.

Well, this is the neurophysiology and biomechanics of it. But 'putting your foot down' in a racing car, at the edge of adhesion, is not quite that simple, is it?

For one thing, this movement, this extension of a single joint, has to be very carefully coordinated in time with other motor commands – which are going through completely different parts of the spinal cord and into the arms and the hands. These other commands are there to ease up on the torque on the steering wheel. Allowing the steering lock to unwind, perhaps even a dab of countersteer.

Also, body core muscles and the neck must be maintained in tension – they need to provide a stable platform for the arms and the head. If the head is flailing about this is not conducive to accurate motor coordination! The eyes and the organs of balance in your ear need a stable platform, too. And eye muscles must keep the thousand-yard stare of the racing driver fixed at the curve exit.

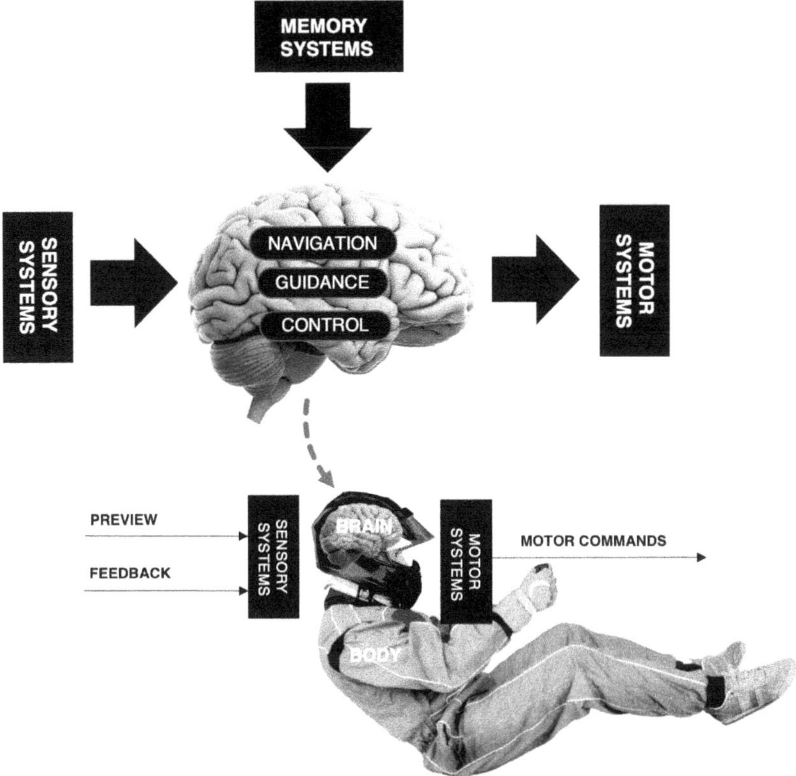

Figure 6-1. The driver's brain is the major locus of control in the driver-vehicle-track system. It takes in feedback and visual preview information through sensory systems, and outputs motor commands through the motor system. The transformation from sensory signals to motor commands is not mere reflex; it is mediated by complex processes based on prior learning and memory. These processes, as we have seen in **Chapter 2** to **Chapter 5,** can be described in terms of control, guidance and navigation operations.

Figure 6-2 (overleaf). Some of the major anatomical structures of the brain. This figure gives you an idea of where some of the parts of the brain mentioned in the text sit inside the head. **Table 6.1.** lists important cognitive functions that depend on specific regions of the cortex.

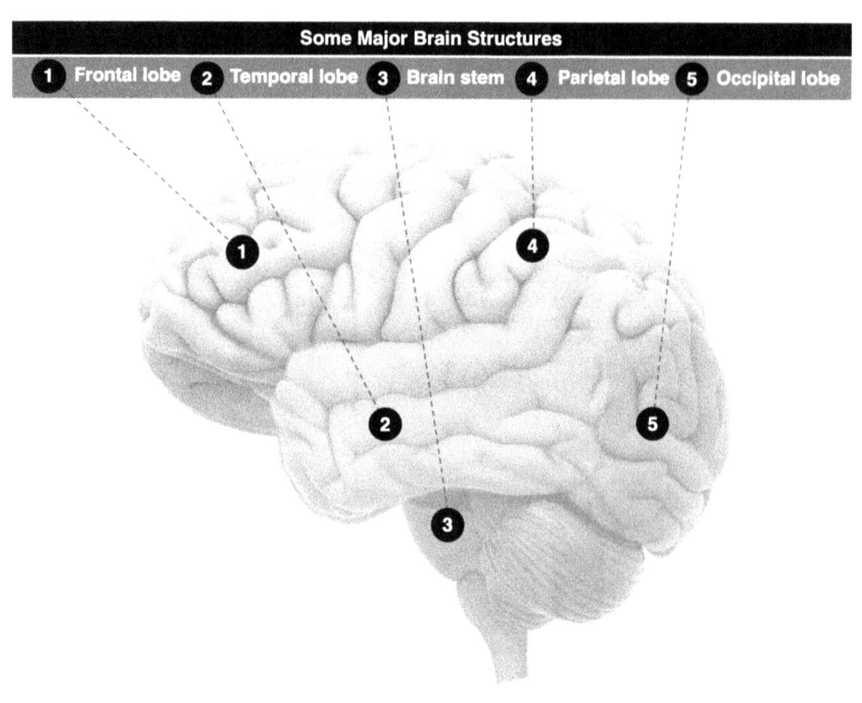
Some Major Brain Structures
1 Frontal lobe
2 Temporal lobe
3 Brain stem
4 Parietal lobe
5 Occipital lobe
1
2
3
4
5

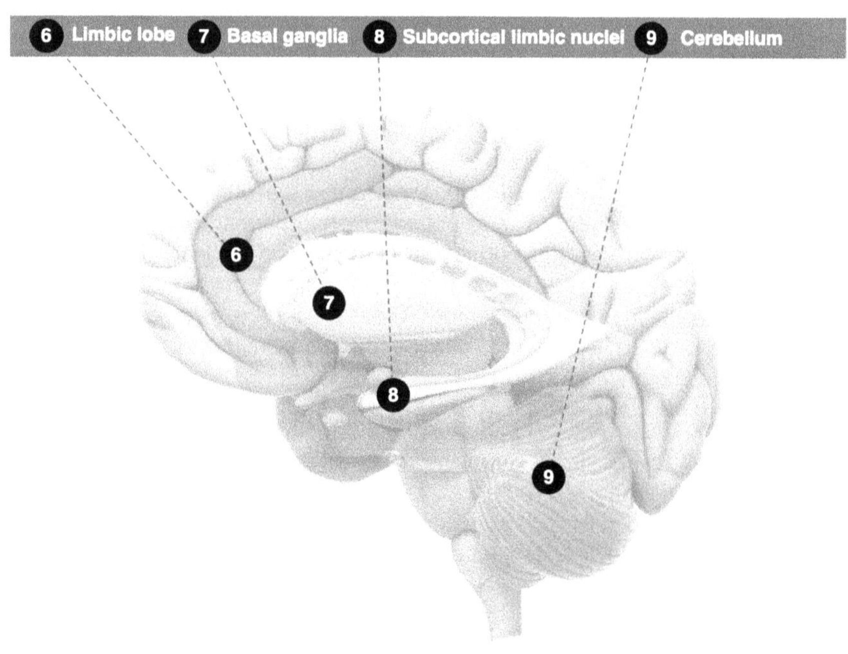
6 Limbic lobe
7 Basal ganglia
8 Subcortical limbic nuclei
9 Cerebellum
6
7
8
9

TABLE 6.1. SOME MAJOR CORTICAL STRUCTURES AND RELATED COGNITIVE FUNCTIONS

Frontal lobe	Temporal lobe	Parietal Lobe	Occipital lobe	Limbic lobe
Motor planning	Auditory perception	Attention	Vision	Emotion & feelings
Executive attention	Object & scene recognition	Gaze control	Object vision *vision for perception*	Processing risk & reward
Procedural memory	Explicit memory	Perception of the body	Spatial vision & visuomotor trans-formations *vision for action*	Spatial and episodic long-term memory
Speech	Language	Sensation of movement in space		

And of course: *exactly by how much* you can start to push the throttle has to be judged (somewhere in the brain) against the perceived speed and stability of the car.

So, at the level of motor coordination this apparently simple act which takes place in a fraction of time – in the end, you're just extending your toes by a few millimeters – already engages all the multiple motor systems of the brain. That is billions of nerve cells, zipping and zapping messages to one another, and to the muscles, all just in order to create a single coordinated action.

And when the brain gets it right, this produces from the vehicle a controlled outcome: an arc through space, the racing line. So, how does the brain produce such control?

REFLEXES ARE NOT ENOUGH

Anatomically, at the level of the spinal cord, feedback is driving reflex responses in a closed-loop manner. Neurons in sensory nerves make contact (synapses) with spinal motor neurons directly, and also via intermediating neurons within the spinal cord. This wiring is behind the clever bit: it is there to make sure that when one set of muscles, such as those for ankle *extension* are activated, the opposing muscles that would *flex* the ankle are not, they are instead relaxed. If they were not being actively inhibited, these muscles would themselves respond, through what's called a stretch reflex, to their being stretched by the ankle movement… and so these muscles would oppose the movement by starting to contract.

This would create spastic rigidity: muscles on both sides of the joint would now be trying to pull it in opposite directions. (This can happen with injury to the spinal cord or the brain.) Any coordinated motor command, even at the reflex response level, means activating one set of muscles, *and* simultaneously de-activating some reflex responses of the opposing muscles.

But even the most automatic and habitual control is not based just on reflexes. Reflexes are closed-loop feedback control, which means they only respond to stimulation that arrives, with some delay, from the sensory signals. Anticipation, and just dealing with the sensory and motor delays in the nerves connecting the brain to the different parts of the body, requires feedforward control of some kind. And this relies on parts of the motor system higher up in the neural hierarchy, deeper inside the brain.

SYNERGIES AND MOTOR PROGRAMS ARE THE BUILDING BLOCKS

The basal ganglia, cerebellum and the brain stem motor structures (**Figure 6-2**) take care of balance, posture and the coordination between muscle groups. They keep the body and head poised. They provide a stable platform for precise and controlled limb (and eye) movements. They twist the neck when you move your eyes to make a glance over the shoulder. They also support coordination of the joints of the wrist, elbow and shoulder, to help you slot in the gear lever in an automated routine action. This is the level of movement *synergies*, that are at a level of coordination higher than reflexes.

Motor synergies differ from reflexes in a few important ways. First, they involve many sources of sensory information, not just a specific stimulus like tapping the knee joint or a flash of light. Second, they involve multiple muscle groups all around the body, and movements of multiple joints, not just a local set of muscles like the ankle extensors and flexors, or the eye muscles. Third, they are largely feed-forward driven; i.e. generated by the brain, based on some internal signals, rather than getting elicited by an external stimulus.

Synergies have been proposed as fundamental building blocks of complex actions.[121] What this means is that specific large movements – e.g. reaching, kicking, stepping… braking, steering, accelerating, …

– are composed by combining together many of these general multi-purpose synergies (and underlying reflexes).

The highest centers of the motor system put together motor programs for skilled routine actions (balancing the car, visual strategy...) and select which of them should be recruited at what time. The procedural memory for learning new skills revolves around learning these new motor programs.

An example within taking a bend could be the sequence of movements for executing a downshift. The motor program is not just about whether to shift down or not – it is also about when exactly to shift, whether to double declutch, and at the highest level also about being able to compensate for how this affects the use of the throttle and steering.

This level of control involves the frontal cerebral cortex (and also special parts of the cerebellum and basal ganglia). The frontal cortex is needed in particular for arranging movements into *(cont'd on page 132)*

Figure 6-3 (next spread, left). Motor hierarchy from reflexes to more complex synergies and motor programs. Coordinated motor action must be integrated across these levels of control: as all levels of the hierarchy feed into the same final common pathway to the muscles, to achieve a coordinated outcome the different types of motor control (distributed all across the brain) must agree on which muscles are to be contracted and which relaxed at any given moment.

Figure 6-4 (next spread, right). The driver's brain must be sensitive to many different sources of information; but at the same time build an unified overall picture of the driving conditions. Perception is the process of sorting and integrating the different signals and combining them with past experience. For this, the brain has to combine signals from the different sense organs (the eyes, the ears and so on). These convey information about the physical environment and the state of the body, but no one single sensory source will contain all the relevant information. What you perceive is the end product of this process of combination, not the physical stimulus energies activating the sensors.

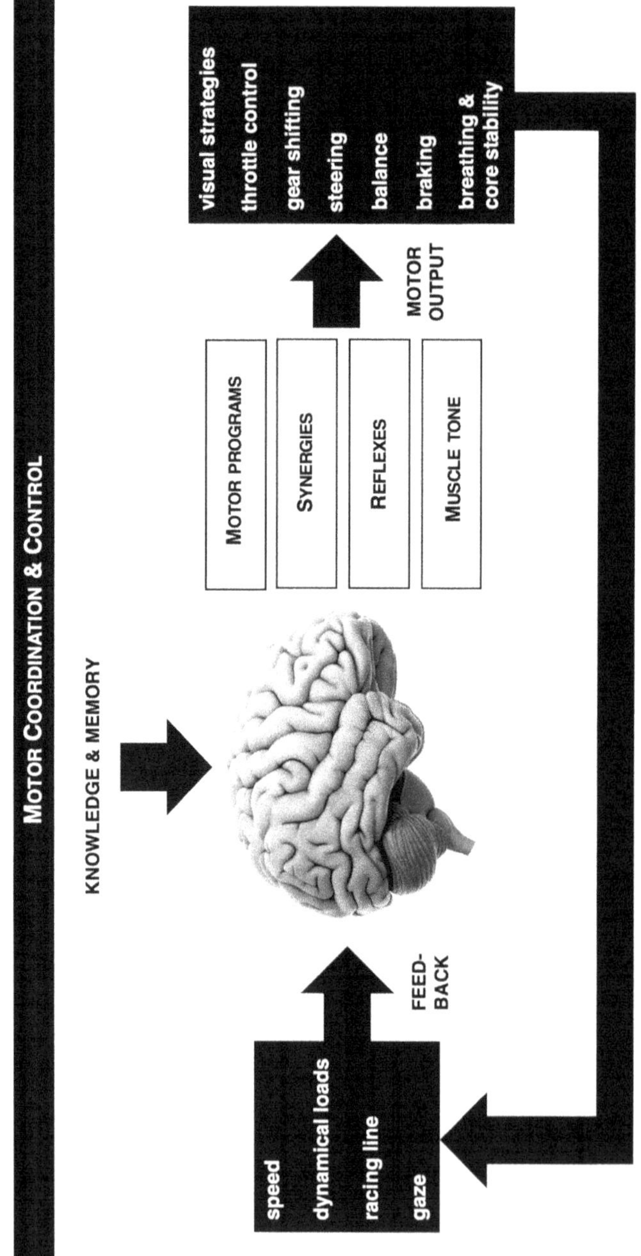
MOTOR COORDINATION & CONTROL
KNOWLEDGE & MEMORY
speed
dynamical loads
racing line
gaze
FEED-BACK
MOTOR PROGRAMS
SYNERGIES
REFLEXES
MUSCLE TONE
MOTOR OUTPUT
visual strategies
throttle control
gear shifting
steering
balance
braking
breathing & core stability

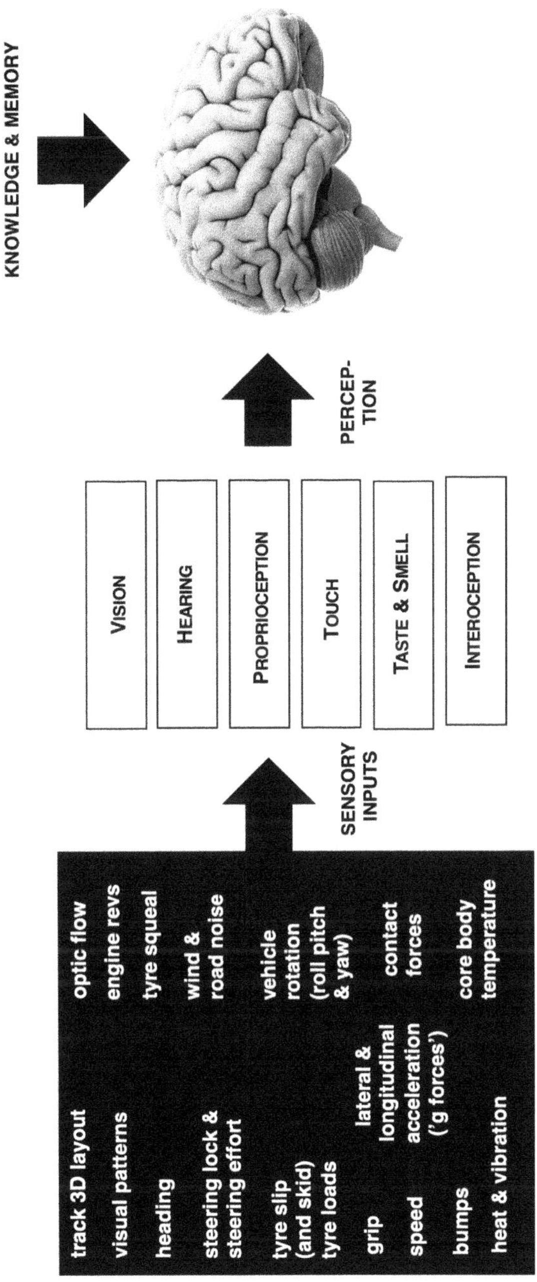
MULTISENSORY INTEGRATION (PERCEPTUAL MODALITIES & FRAMES OF REFERENCE)
KNOWLEDGE & MEMORY
PERCEP-
TION
VISION
HEARING
PROPRIOCEPTION
TOUCH
TASTE & SMELL
INTEROCEPTION
SENSORY
INPUTS
track 3D layout
optic flow
visual patterns
engine revs
heading
tyre squeal
steering lock & steering effort
wind & road noise
tyre slip (and skid) tyre loads
vehicle rotation (roll pitch & yaw)
lateral & longitudinal acceleration ('g forces')
grip
contact forces
speed
bumps
core body temperature
heat & vibration

(cont'd from page 129) more complicated sequences, sometimes called 'kinetic melodies'.

It is also needed for making sure that they are appropriate for the context. If you reach *for* something, kick *at* a ball, or accelerate *through* a particular curve, then the movement needs to be organized at this motor program level above synergies and reflexes. Anatomically, above the spinal cord/brain stem level.

So, with this knowledge, think again of our driver. In our super-slow-motion image he is still where we left him: 'balancing' the car on the throttle. He is making it accelerate and rotate at the same time, through minor adjustments of the throttle and the steering angle. Remember that it is no use for him to apply the throttle 'perfectly', if the steering is not released at the same time with the appropriate rate, as determined by the dynamics of the car, and the available grip. It is no use to unwind steering 'perfectly' if he is not hard enough, or is too hard, on the throttle. In fact, there is no such thing as 'perfect' throttle application except in the context of current steering. There *is* no 'perfect' unwinding of the steering, except in the context of the current throttle application. (And dynamic load distribution and grip and myriad other variables we are not considering here.)

Everything is *interconnected* in the vehicle's dynamics. That's why it needs to be *interconnected* in the brain also. This is what well-integrated motor coordination, from reflex to the highest levels of the motor system, will give you.

DEALING WITH DELAYS

The synergies and motor programs depend on the brain's in-built and learned wiring, and an expert athlete will have a huge number of motor programs to call upon. But how does the brain know which motor programs should be activated when?

To be able to coordinate the ankle extension (throttle) against shoulder rotation and the pressure of the palm (steering) against visual and vestibular inner ear feedback (speed and rate of rotation), the brain must not only coordinate action across multiple muscle groups and joints. The brain must also combine different forms of feedback and preview information from multiple sensory modalities. (**Figure 6-4**, previous spread).

Information carried by signals along different sensory nerves arrives in the brain at very different forms (the sensory organs measure events in different coordinate systems), and at different delays. Neuroscientists call the process of combining all these signals 'multisensory integration'. Psychologists call the same process perception.

Feedback received through the senses does not just drive reflexes. It is also sent up sensory nerves and along the spinal cord deep into the brain. There the feedback from the multiple sensory systems will be compared, sorted and interpreted in light of expectations and memory. That is: information already residing in the brain.

Because of delays in neural and synaptic transmission, feedback will arrive in the brain after some not at all inconsiderable latency (0.1 to 0.2 seconds). So, does this mean that the brain is inevitably 'behind' what is happening? It is indeed sometimes claimed that because of sensory lag and perceptual processing time, we live about a half a second in the past… But no. Because the sensory information coming in right now is just part of the story.

The brain has a powerful trick up its sleeve, when it comes to integrating all the available information into perception and action. Because it is not a simple reflex-engine, the brain can *anticipate* events.

The racing driver's brain can keep ahead of the vehicle: as long as the vehicle is behaving *predictably*, the countersteer of the elite driver who is 'on top of the car' can begin with zero reaction time – or even negative reaction time. The driver can start correcting before the car itself even knows it's going into oversteer. That's because the brain can rely on the information from prior learning to predict and to control things.

Simply responding by reflex to what the car is doing, perceiving it with latency and only then responding, the driver would indeed be constantly 'behind' the car: compensating, overshooting, recompensating… (This is exactly the problem why in engineering closed-loop controllers must be supplemented by feedforward control).

It can happen, especially to the inexperienced driver. Even for an experienced driver this can happen when the advance information (the internal models the sensory systems feed into, more of these in a minute) can no longer stay 'on top of the situation'. When the brain falls behind and cannot issue the right anticipatory motor commands. For example, in the oscillatory oversteer condition known as a tank slapper. This very often leads to a spin, i.e. loss of control.

To avoid this happening (too often at least) the brain must be using powerful feedforward control. In neuroscience jargon, the racer's brain is said to encode internal models of the car's dynamics. ('IF the car is doing this AND I do that, THEN how would it respond?', 'IF the car is doing this AND I want it to do that, THEN how must I operate the controls?') This is what allows the myriad variables that need to be judged to be lined up and weighed against one another. All this happens a split second, and mostly unconsciously of course. You are only aware of the *outcome* of this process, not the raw input or the intermediate computational steps.

If the brain were to wait for the feedback to arrive, and *then* correct the steering, only reacting to the *effects* of throttle application, it would be too late. Given the physiological delays in neural transmission, and the time it takes brain cells to receive and analyze signals, the brain would be always behind the car. But if it is running a kind of 'internal simulation' of the car's dynamics, then the brain can access the information in its simulation *right now.* (The idea of a 'simulation' of the external world is a metaphor, not a 100% accurate description of what an internal model does, but gives you the idea.)

So, you can see how, from a brain design point of view, there is a big difference between *reflex* actions proper, and *automatic* actions or *habit.* All reflexes are automatic actions, but not all automatic actions are mere reflexes. In skilled behaviours, most of them are actually anticipatory, feedforward actions based on internal models.

NEUROPHYSIOLOGICAL ORIGINS OF 'SMOOTH' CONTROL

Racing skill is a balancing act – not just balancing the car, but balancing many muscle groups and balancing multiple sources of feedback. Next, we want to talk about how this fact speaks to the question of 'smooth' control, and different driving styles.

'Smoothness' is something that all high-performance driving instructors agree is somehow important for speed. But what is 'smooth', actually?

Ideally, according to many driving textbooks, throttle application needs to be done in one movement. You get on the throttle, and progressively, 'smoothly', increase it until you are on full power by the exit of the corner. If you have to back off and then get back on the throttle again it means you are probably doing something wrong. And you don't release the brakes suddenly: you 'smoothly' disengage them. A 'smooth' driver is not continually responding to incipient skids and sawing on the steering wheel – she 'smoothly' turns the car in and then unwinds the steering, all in one elegant motion…

This way of describing it *almost* suggests a motor program that has been set up in advance, with precise parameters, which then simply gets launched at the right moment. But as we saw earlier, this is not the way of the brain. The brain is organic. It is not a machine. For a human, metronomic precision is not the way to achieve highest speeds and performance.

From the brain's point of view, this *must* be so because the brain simply *cannot* know the exact parameters of very complex movements in advance. It has in memory a lot of information about the world, sure… but not a perfect replica of the world (the world is a big place). The brain does not have to rely on just feedback because it can query its own internal forward model, yes… but this model can never be complete, totally accurate and without noise and error.[122]

If you are really going fast enough, there is always some element of uncertainty to what the car is going to do. So, does the racing driver need to be going slow enough for the brain to be able to pick up and use feedback (in closed-loop control)? So that the brain can receive, process, and compensate for uncertainty in the motor program…? No. That is no good either. You want to go fast!

So, it now seems that as a strict objective simply to 'be smooth' doesn't seem to make sense after all? You can certainly drive with less uncertainty, less variability in inputs, more 'smooth'… by going slower. But you do not want to be 'smooth' by going slower – you want to be 'smooth' but at the same time fast and decisive! Paradox!

Here is what I (OL) think might be going on. Consider, for example, shifting gears. You should be decisive and move the gear lever fast and with intent… but if you are rough and hurried, that is not smooth. Smooth implies finesse. You should move the lever quickly – but not faster than what the 'box will take. Similarly, the wheel should be turned quickly and sharply – but not so fast and so much that the slip angles drop off the peak of the slip-grip curve. A little bit slower, or a little less peaky build-up to the speed of moving the controller, and the tyres and chassis have time to respond. So, perhaps one could say smooth is about being quick and decisive with the controls… but in tune with how fast the system being controlled is actually able and willing to respond.

Also, when working with drivers in practice, it strikes me (AD) how for every Jim Clark there's a Colin McRae. For every Rossi there's a Marquez. Whether the 'style' of a driver or a rider is smooth and economic or a 'wilder' style with more variability in inputs is partly a matter of individual preference. Jenson Button was always a smooth driver (ever since he was in karts), and this never really changed all throughout his career. Or google up Senna vs. Prost at Paris-Bercy 1993: despite the massively different nature of karts to F1, they still display their fundamental and individual driving qualities in each (very different) machine.

In fact, Ayrton Senna had a fascinating throttle technique. Multiple small applications at a high frequency[123] – these are not responses to errors, they are part of the strategy, part of the motor program – and if you try it, it doesn't destabilize the car, it kind of feels like traction control. But there's a world of difference between 5-6 fast stabs of the throttle, and 1-2 big lazy ones. The same thing with steering. Michael Schumacher was renowned for micro-adjustments on the wheel.

And this, I (OL) think is the key to cracking 'the paradox of smooth': it is ok to make very fast inputs in succession, if they are gentle and small enough that the effect of each will be 'smoothed out' by the dynamics of the vehicle. In technical terms, the vehicle's inherent dynamics will low-pass filter movements that are of a small amplitude and above the response frequency of the vehicle – i.e. the vehicle smooths the inputs in producing the output (in signal analysis, another word used for low-pass filtering is, indeed, 'smoothing'!)

The speed and the driving line and the load transfers are the invariant outcomes that the brain is really controlling, remember. These are the true outputs – not the muscle contractions, or even the movements of the controllers. It's just like the blacksmith controlling the tip of the hammer, not the grip and the arm-angles as such. So, variability and 'smoothness' at the level of the outcome is what matters to the brain, more than the 'smoothness' of the motor actions or muscle commands that produce that outcome.

Not one person I (AD) have come across actually teaches the Senna-type technique, though.[124] In fact, I've only found one other person who has the same technique. He happens to be a sim driver from America who developed it when he had to use a keyboard as a throttle. Sharp high frequency throttle inputs. My guess is that maybe when

Senna started karting, his throttle, for whatever reason, wasn't particularly progressive and thus he incorporated this strategy into his driving style. Some attribute it to him driving the 80s F1 Turbo cars and that it helped spool the turbo, but having seen him do it in a kart prior to that, I don't think it came from the turbo cars.

While trying to figure out the concept, I had to experiment with my own driving. The peculiar thing was that I do actually use this technique when the car is in oversteer. I can't tell you why I do it or what effect it has, it's automatic. One of the odd things is I would approach a corner and try to imitate Senna's throttle, struggling to match the frequency and get a good feel for it. But if the car suddenly snapped into oversteer then I would automatically start pulsing the throttle. I can't begin to tell you how peculiar all this is for me to understand. I can't imitate this technique under normal conditions, it doesn't feel right – but then in a split second I am doing this weird technique to rescue a slide. I can't help but feel it has some significance on how the brain interprets different sources of information…

So, it looks like next we do need to understand more about the actual processes for obtaining sensory information.

PERCEPTION COMES IN DIFFERENT FLAVORS

Let us look more closely at the information that is impinging on the sensory organs of a driver. There are the patterns of light on the retina as the racetrack whizzes by. (Scientists call this optic flow, where the visual world seems to flow by a moving observer.) The sounds of the engine and tyre scrub will reach the ears, despite helmet, balaclava and earplugs. The mechanical forces on the palms of the hands create signals about the ebb and flow of steering wheel torque, and these are sensed through the skin; as are the movements and vibrations of the seat against the driver's backside. Lateral g-force is registered in the vestibular organs embedded inside the skull bones. And muscle tension and joint angles, as the driver works the wheel and pedals, are sensed by stretch receptors within the skeletomuscular system. (Refer here back to **Figure 6-4**.)

These different stimuli are all recorded by various types of tiny sensory cells, each specialized to respond to their specific type of physical stimulation (light, pressure, vibration etc.). The sensory cells are connected to nerve cells, neurons, specialized in transmitting signals along nerve fibers. These signals are discrete pulses, called action potentials, whizzing into the brain at the speed of some 100 m/s and a rate of about 1 to 100 Hz. If you insert a small electrode into a sensory nerve and hook it up to an amplifier, it sounds like a Geiger counter.

Six senses give the brain access to the world: vision, hearing, smell, taste, touch and proprioception. That 'sixth sense' of proprioception refers to sensory systems that give you a sense of your body, a sense of your body parts belonging to the body, and a sense of your body in space. Posture, body movement and position, limb and joint angles, muscle tension, and all the motions of the body parts relative to one another are registered by sensors that feed into the proprioceptive systems of the brain. Also, orientation relative to gravity, which is important for your sense of balance, can be considered a form of proprioception. Then there's the sensation of pain, heat and cold; this too is part of proprioception.

(Proprioception is not any more mysterious than the five usual senses, it just did not have a name in the ancient catalogues of five senses, so scientists had to give it a weird name. Just saying – to make sure nobody gets funny ideas about racers using 'a sixth sense'.)

Then there is interoception, which is sometimes considered a separate sense. This is for things like core body temperature, hydration and blood sugar levels etc. Interoception tells your brain how your body is doing. Pain in the gut or a headache can also be considered interoception. (Though we tend to call these experiences perhaps more often 'feelings' than 'perceptions'.) So, if you count this as a separate sensory system then there are actually seven senses.

More important than exactly how many different senses there are, and more important than the specific categories and under which latin or greek labels you choose to classify them, and more important than what their exact anatomical sense organs are like, is the following general principle: the brain, floating in the dark cavity of the skull, only has access to the outer world via these separate streams of nerve signals, responding to different things, and coming in at different latencies.

For the brain, this is a big, big challenge; the information about any relevant physical event – such as oversteer vs. understeer, or other important things in life – is not contained in any one signal. But by cleverly comparing the different signals to one another, the brain can patch together a coherent overall picture of the situation. This process is called multisensory integration, and the resulting picture is called perception (or an internal model).

Take, for example, the perception of speed. This is patched together from visual flow on the retina and the sounds of the tyres, the engine and the wind, and from accelerations sensed through the vestibular organs and proprioception. Then there is the perception of grip, from the feel of the steering wheel, tyre squeal, vehicle yaw responses etc. These will *all* come together in the apparently simple and direct feeling of *'Boy, this feels fast!'*.

What makes this process particularly impressive is that all these signals encode different aspects of the situation in quite different coordinate systems,[125] and arrive at a latency up to a few tenths of a second (in a race car a lot can happen within a few tenths of a second). And on top of that, the latencies of the different sensors are not even the same.

Auditory signals, for example, are encoded much faster than visual signals, because the sensory cells of the ear register pressure changes much quicker than the retina registers changes in pattern of light. And it takes a measurably longer time for touch signals to arrive from the toe than from the shoulder, simply because of the extra distance.

So, integrating the distinct sensations into a coherent, unified, perception really requires much more sophisticated machinery than you might think. It might seem like a pretty simple and straightforward process, given how immediate our sensory experience and perception feels like. But really, underneath, it is anything but. You've got billions and billions of nerve cells dedicated to it. And the brain's computations are amazingly reliable and efficient in it!

VISUAL GUIDANCE

You could argue that the most important sensory organs for the racing driver are the eyes. These give the necessary information about road geometry ahead. Yet as we've seen, on the basis of conscious introspection we have remarkably limited insight into how high-speed visual guidance works. So, let's analyze a bit what is happening here.

Moving at high speed critically depends on visual preview. Avoiding obstacles, anticipating changes in heading, predicting the future trajectory – for all these purposes vision is the only sensory sys-

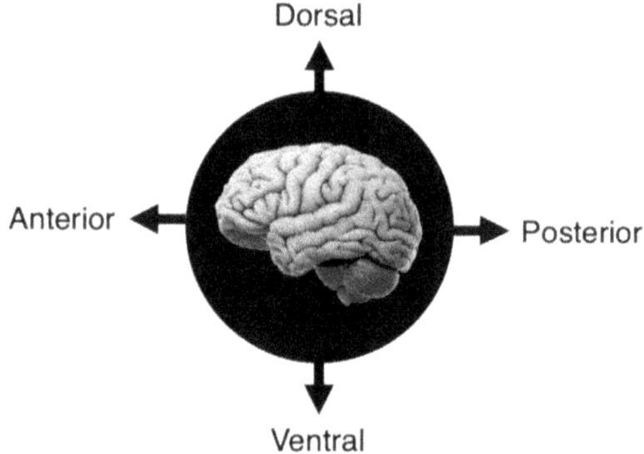

Figure 6-5. The names of the cardinal directions neuroscientists use to refer to parts of the brain.

tem that can give us detailed information about the scene layout at really far distances. You can feel a bump on the road through your backside, but if you see it ahead in visual preview you can avoid it altogether.

(Also hearing can give humans surprisingly accurate information about the space we are in, but for humans – lacking the sonar capability of bats – this is not detailed enough to guide high-speed locomotion. And in a racecar the sound of the engine, tyres and wind of course pretty much drown out everything else.)

For a driver, it is vision that allows him to perceive the curvature and elevation changes and to judge distances – the physical scene layout delimiting how the car may be driven. Reference points are of course recognized by landmark features.[126] Also, the speed and the rate of rotation of the car is registered visually, from optic flow.

Besides road geometry, surface adhesion makes a big difference to how hard you can steer and accelerate. This also can, to some extent, be gauged by visual preview. Rally drivers have to develop this skill to judge ice situations on the Monte, for example – here the brain can't rely on memory from the previous laps. The texture, colour and sheen of the surface are visual clues about the grip on offer ahead.

The analysis of fine texture is handled by its own set of higher-order visual processors. But it is also proprioception and feel through the palms of your hands that allow you to 'read' the grip levels, partly supported by the visual information. These signals have to be integrated across different systems. All this is analyzed in many, many

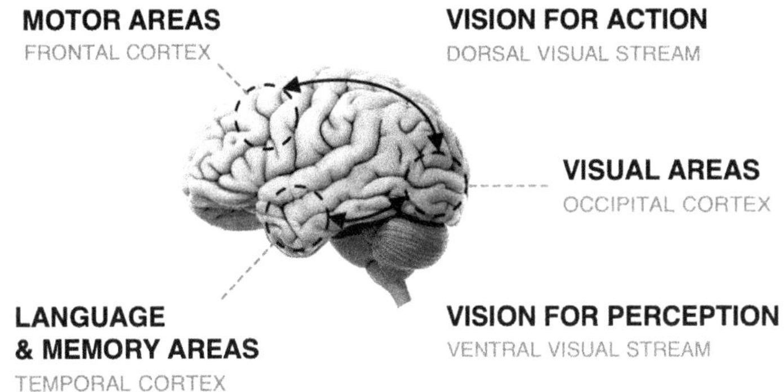

Figure 6-6. Dorsal and ventral visual streams connect the visual cortex to motor areas in the frontal cortex, and language/memory areas in the temporal cortex, respectively. They are sometimes called the vision-for-action and vision-for-perception systems.

visual cortical areas and there is no single 'master visual image' integrating everything in one place (just like there is no 'master controller' on the motor side, by the way).

The strange thing is: dealing with the delays and different coordinate systems of different body parts does not arise only in sensory analysis and motor control, exactly the same kind of coordination problem presents itself inside the brain as well.[127] You see, the road-geometry-to-motor-commands transformations (vision-to-action operations) are handled by one visual system, and the analysis of texture, colour and sheen, and the sensation of movement in the visual field (vision-to-perception processes) are handled by quite different brain systems. Even though both are what we would call vision, the systems are working on different visual analysis problems, at different time rates.

Anatomically the two systems are called the dorsal and ventral visual streams. They extend from the primary visual cortex, which is the first waystation of visual signals in the cerebral cortex, towards different higher-order association cortices.

(Dorsal, by the way, in the neurological coordinate system means on the top-side and ventral means on the bottom-side; the same uses as in 'dorsal fin'. Refer to **Figure 6-5** and **Figure 6-6**.)

Although both streams work with optical information, they do different things. The dorsal stream links visual information to motor planning. Generally speaking, it extracts from optical signals 'where and how?' information, about directions and distances. The ventral stream links vision to memory and category information: 'what?' objects and surface textures and colours there are. So, whereas the motor system must *converge* on a final common pathway to muscles, the perceptual analysis can and does *diverge* from the primary sensory waystations (as well as converging for multisensory integration).

EFFERENCE COPY – A NON-SENSORY INPUT TO PERCEPTION

Driving a racing car is a full-technicolor surround-sound experience. From the brain's point of view, this means multisensory integration. The grip of the road surface is not only predicted on the basis of visual information ('How does that patch of asphalt look?'), but these predictions are constantly checked by feedback from the steering wheel and seat ('How do the contact forces and g-forces feel?').

But, remarkably, the story of perception does not end with multi*sensory* integration. There is another major input that does not come from the sensory organs, but from inside the brain itself: motor commands sent to the muscles are also sent to the perceptual system for integration. Neuroscientists call these signals efference copy.

Why would *motor* signals be sent to the sensory brain areas? Doesn't this mess everything up? To the contrary. Efference copy is used by the brain to tell which changes in the sensory signals are caused by events out in the world, and which result from moving the sensory organ, or the body!

This has been studied extensively with eye movements, my (OL) specialty. So, please indulge me in running through an example in a bit more detail.

Suppose our driver is on a curved trajectory; you can analyze this as a combination of forward motion at some speed and yaw at some rate of rotation. The driver may be also visually tracking an object in the distance, say, the exit point of the turn, as it sweeps across the visual field (when the car rotates in yaw to the right, the target sweeps to the left). This gaze control is called a tracking fixation, or pursuit eye movement.

We already covered the multisensory perception of speed (from optic flow, sounds, vibration etc.). To recap: keeping oneself oriented in space, whilst rotating, is dependent on acceleration-sensing organs in the inner ear that register the combined effect of gravity and g-forces on the car, and yaw rotation is additionally sensed by the eyes, as the visual world spins around you.

But the eyes and the head are not spinning passively. The head and the eye counter-rotate in the opposite direction, to (partially) stabilize the visual image sweeping across the retina. How?

On the motor side, there is first of all an automatic reflex-response to counter-rotate the eyes. This is driven bottom-up on the basis of the vestibular stimulation and the visual flow of optical texture on the retina. Neuroscientists call these reflexes the vestibulo-ocular reflex and optokinetic response, respectively. They compensate the blur of the optical image on the retina. (Not completely, there is some residual blurring). There is also top-down tracking, driven by the internal model of eye-head-body-vehicle movements.

Now, when the brain tries to make the most sensitive and accurate estimate of the actual yaw and velocity (of the eye-head-body-vehicle) that it can, it will combine all these sources of information: visual, vestibular… but also motor. Because from efference copy the perceptual system finds out how the motor system is currently telling the eyes and the rest to move.

Your brain is effectively going: 'there is X amount of visual motion reported by the eye… but wait a minute, the motor system is turning the eye by Y – so I must be actually rotating by X + Y… and… hey, what do you know… the vestibular system is actually saying I am rotating by Z, and Z is almost exactly X + Y! So that must be it – it all fits!'.

This is not happening in your conscious mind, of course; but some sort of 'unconscious inference' like this is nevertheless keeping your internal models up-to-date, on a millisecond-by-millisecond basis.

What the inner ear organs of balance are saying (Z), what the retinal slip of the image sweeping across the eye is (X), *and* what motor commands are being given to the eye (Y) are all kept in register relative to one another. That this happens within milliseconds with slow and noisy processing units is pretty clever biological engineering… you just gotta love it…

The main point here is that perception is not just about registering sensory information. Efference copy of motor control is used in perception itself. Only those movements of the image on the eye that were *not* caused by a motor command end up being perceived as movement 'out there'. (Remember looking at yourself in the mirror and failing to see your own saccades?)

Two identical retinal image movements with identical patterns of sensory inflow can be perceived in different ways; either as movement, or as a stable scene if the brain knows that it has itself generated all the image motion by moving the eye. More precisely: if the visual system can unconsciously infer this from efference copy. (You don't experience 'lights-out' when blinking.) This is why we experience a stable visual world, even when our eyes flit around many times a second.

So, do not be surprised when you look at the brain wiring and find that the motor system does not only send motor commands to the muscles, but that it also sends 'motor commands' to the sensory systems! The brain is not a one-way street.

All these different sources of information complement – or sometimes conflict – one another in perception. Only by combining all of it together can a full picture of the driving conditions be patched together. All millisecond-level sensorimotor stuff, deep, deep, below the level of conscious awareness; it is the outcome of this automatic, multisensory combination process that we will be aware of.

'FEEL' (REPRISE)

And here we are again, with our racing driver. Now exiting the final corner, committed on the throttle, easing off on the steering, and feeding in the power. Small mini-corrections of opposite lock are managing the rotation of the car, almost anticipating its reactions… 'smoothly'.

The loaded outside rear wheels are rotating the car towards the inside of the bend, steering into the bend. (If the rear wheels were not steering, they would be sliding in a straight line and the car would be rotating even faster, in fact skidding and then spinning.) The unloaded inside front wheel almost lifts off the ground.

The optical stimuli are whizzing by, the g-forces are being registered in the head, as is the pressure of the outside edge of the seat that is pushing to the shoulders, the back and the thighs (accelerating the driver's body towards the inside of the bend, and making the body take the same curved path in space as the vehicle). This information combines into a perception of speed and a proprioceptive sensation of lateral load.

The contact forces on the palms through the steering wheel are ebbing and flowing, signaling the grip level. Through the sole of the right foot, pressure on the throttle pedal is sensed and combined with engine note and rate of rotation to judge traction… these motions and forces are registered and integrated into the 'feel' of the car, and its 'response'.

But now consider this: any novice driver is in possession of the very same sensory organs. So, in theory she has available all the same sensory information. Yet a novice cannot *feel* the car in anything like the same clarity and detail an expert can. Why is that?

The reason is that what we experience as the world 'out there', e.g. 'the feel of the car', is not simply a product of encoding stimulus information originating in the external world, coming through the senses (plus efference copy). So far, we have considered the brain as mostly having access to the stream of sensory signals and generating motor signals sent to the muscles. But there is one more, vast and remarkable, body of information. All the information in the brain itself, which has accumulated into memory from past experience.

Consider this: the sensory information is generated during one split second; the memory information could be learned over a lifetime. So, in terms of the amount of information available, which do you reckon is going to be the bigger of the two?

As we saw in earlier chapters: in general, sensory signals only make sense if the brain can relate them to some information already stored in memory – chunk them, associate them… So, in feeling the car the racer's brain integrates all the inputs and outputs to memory information as well.

It's the same situation in any field of expertise – be it cooking, wine tasting, fine art, or rock climbing, or indeed race driving. The connoisseur can resolve the situation in a fine level of detail, in some cases so much so that it is as if they lived in a different world to the rest of us.

To properly understand these processes we need to take a look at some more advanced brain theory.

Chapter Seven

Analyzing the Complex Brain

WHAT ARE THE MEMORY PATTERNS that our driver has stored in his head? What are they like, physically? How does the brain shape its own tissues to store information? Where are internal models and feedforward motor programs engrained in the living brain? And how did the *right* information get stored in the first place?

To begin to answer such questions, one needs to develop some general concepts about how the brain works. Specifically, how the brain and its processes are organized into *complex functional systems*, and how these relate to the *localization* of different cognitive processes.

What we have here is material needed to interpret, in a scientifically valid way, practically all findings from any brain imaging research you will ever have come across, or are likely to come across. All those books and magazine articles with colourful blobs painted on the brain.

If you do not at some point go through the trouble of building up the correct thinking habits, you are liable to misinterpret a lot of it. Specifically, a system-level overview and an intuition of how the actual logic of defining 'brain activation' works, is needed to avoid mapping 'psychological processes' directly into brain areas in a way that does not reflect how the brain is put together. (Like: 'here is the steering cortex', or 'this is the car-balancing nucleus' and 'that's the sponsor-schmoozing area, right next to the excuses-to-team-manager region…').

While it is often interesting to find out about 'activation' of specific brain regions, or about differences in local brain structure between different people – both things we will be getting into – it's also important to get an idea of what activation actually means, and what those anatomical differences can actually tell you and what they can't.

This may all be a bit abstract and difficult to wrap your head around at first. Sorry. These are big topics, and not easy to get straight. So, feel very free on first reading not to get bogged down on the details, and just skim along to get the gist.

If you've seen colourful brain images with blobs of activation, or read pieces about brain scientists finding 'the brain center for X' (for X insert just about any psychological faculty), you may have been tempted to start developing exactly the wrong kinds of intuitions, without even noticing. Those we're now going to ask you to confront and challenge. Fortunately, racing technology gives you a very good and valid set of alternative intuitions!

Because both the brain and the racing car are complex functional systems, being able to think about the behaviour of a racing car already gives you one concrete example of how to think straight about such systems. If you have a 'systems-level' understanding of a racing car, you are already better set to develop a systems-level understanding of the mind and brain than most people.

(We are not talking about having a degree in engineering here – if you have a real-world feel for how the thing is put together, you will have developed the right sorts of intuitions. Those we ask you to build on.)

To be concrete: think of the 'function' of steering a racing car. Then consider the differences in 'localized activation' in various components of the car. Take, for example, the steering wheel. *Localized activation of the steering wheel* means that sometimes it's turning to and fro, and sometimes it's still. If you were to 'image' steering wheel activity (i.e. look at a telemetry trace) you would find that on bendy roads there is a lot of activity and on more straight roads less activity. So, you have found localized activation that responds to different types of road! Great.

Now, obviously the steering wheel is part of the steering system. But you also know that *steering* is a complex, nuanced, skillful car-control process; it cannot be localized just into the behaviour of the steering wheel. The car does not steer whenever and only when the steering wheel rotates, and *the local patterns of activation in the steering wheel telemetry trace do not contain enough information to understand what is going on.*

For one thing, the steering wheel obviously needs to be properly connected to the tyres via the steering column. Otherwise the car will experience what neurologists call a dissociation syndrome – that's when the parts that are needed to carry out some function are working normally, but not properly linked. Further than that, the car needs to be in motion, and the wheels rolling on a surface with adhesion. You may turn the wheel while stationary or on ice, and the front wheels turn… but the car does not steer!

And if you know your vehicle dynamics,[128] you will know that it is not only the front wheels that do the steering. The front wheels are of course mechanically linked to the steering wheel, and they are the ones that you can turn. But it's the slip angles of all four wheels that steer the car. And since the steering effect of each tyre depends not only on the slip angle, but on its vertical load, it follows that the brake and throttle pedal – which you use partly to control the distribution of those loads – also are part of 'the steering process'. In neuroimaging jargon we would say that 'the circuitry that is responsible for the function of steering is distributed across the car'. Every racer instinctively knows this, and terms such as 'steering with the throttle' reflect this.

Now, the 'visual cortex' is to vision what the steering wheel is to steering. Necessary, but not sufficient. Part of a distributed network that is jointly responsible for the function of vision. It has a specific role in that function, but it does not produce vision on its own.

You cannot *localize* steering as a whole to just the steering wheel. Yet, clearly, the steering wheel does something definable and definite for the steering process (it rotates the front wheel relative to the chassis). It's just that this specific contribution, this operation, is not the same thing as 'steering as a whole'.

You cannot *understand* understeer, oversteer, steering feel etc. by a careful analysis of the componentry of the steering wheel, and the way it 'lights up' in a rear-wheel skid! You need to look at the whole system in its context (the movement of the car) to define what understeer, oversteer and so on are.

Likewise, you cannot *localize* vision just to the visual cortex. The operations the visual cortex does is not the same thing as vision as a whole. It does have specific contributions, but you cannot *understand* vision just by looking at its componentry and the way it 'lights up' in a visual task. You need to look at the whole system in its context (patterns of light).

It's not that steering, or vision for that matter, is such a mysterious and magical process that it can *only* be considered 'wholistically', or that it would be distributed across *every* component of the whole car. (The fuel cap for example is not part of the steering system.) And there *is* a basic operation (turning the wheels, changing slip angles) that you *can* in fact localize to the steering wheel and the steering column and the contact patches, a specific 'circuit'.

The subtlety of the logic here is that only when you are able to analyze the function – to take 'steering' apart, logically – will you come to understand it and the components' roles. The same is true with any psychological processes in the brain: you need to be able to analyze complex cognitive processes into operations that are simple enough to localize in specific circuits. That way you can relate elementary functions of the mind to elementary components in the brain. Just looking at the components on their own, or their 'activations' (or patterns of 'co-activation' even), does not give you this, though.

Now, unfortunately the problem brain imaging faces is a bit like trying to infer the role of the steering wheel from telemetry traces for the steering wheel on different roads... *without* having a theory of vehicle dynamics! (Bummer.)

After all, a theory of the brain is what neuroscience is hoping to achieve some day. The very reason we want to do brain imaging in the first place is to learn more about the cognitive processes underpinning the mind, brain and behaviour... chicken and egg come to mind.

There are, however, some general design principles of the brain that are useful. They are not an exact theory, but they are principles that you can apply to pretty much *all* cognitive brain functions to take them apart, logically.

UNDER THE HOOD – SOME GENERAL 'LAWS' OF BRAIN FUNCTION

We will once again organize our discussion around fundamental 'laws'. This time there are five. Five general laws of brain function listed in **Table 7.1.** (overleaf). Let's look at each of them in turn. (First an overall tour of the big ideas; we will go through examples in race driving in the next chapter.)

TABLE 7.1. GENERAL 'LAWS' OF BRAIN FUNCTION

1. Memory is cheap (and fast)	
2. Thinking is slow (and expensive)	
3. Hierarchical organization	3.1. Neuroanatomical hierarchy
	3.2. Information processing is organized hierarchically
4. Wired for prediction	4.1. Bidirectional connectivity
	4.2. Predictive coding
5. Learning from prediction error	

1. Memory is cheap (and fast)

2. Thinking is slow (and expensive)

What the brain likes to do, what it's really efficient at, is matching current perceptions to stored patterns in long-term memory. Recognizing complex situations as a whole, top-down, rather than building a picture of the situation from separately identified features bottom-up.

This relates directly to our discussion in **Chapter 3** on how expert performance relies on pattern recognition. The retrieval of knowledge from a vast store, in a split second, is what the brain excels at. This ability underlies the fast and automatic performance needed in all areas of expertise studied, especially so in high-speed skills like race driving. It gives the chunking processes their (misleading) appearance of immediacy and simplicity.

In contrast, building up new chunks, new patterns, and consolidating them into stable long-term memories is slower (by an order of magnitude at least – minutes to hours as opposed to split-second).

3. Hierarchical organization

Brain systems are hierarchical. This applies to both information processing and to neuroanatomical connectivity. Higher levels of abstraction are processed at higher (deeper) levels in the neural hierarchy.

Form follows function. This is, by the way, one reason why we chose to analyze the task of the racing driver in terms of a *hierarchy* from control to guidance to navigation, rather than, say, the *sequence* from braking to turning to acceleration. A hierarchical analysis of the task is more natural to align with brain function, because the brain itself is hierarchically organized. So this perspective automatically makes our job description, our cognitive analysis of the task, more brain compatible.

3.1. Neuroanatomical hierarchy

The perceptual and memory systems in the posterior (rear) brain and the motor systems in the anterior (front) parts of the brain are organized into hierarchies (**Figure 7-1**, overleaf). Here the levels are defined by counting the number of synapses from the sensory and motor cells.

The sensorimotor periphery (areas directly connected to sensory organs and muscles) is the 'bottom'. The more synapses you need to traverse to reach a structure deeper in the central nervous system, the higher that structure is in the anatomical hierarchy.

Sensory and motor neurons and interneurons in the spinal cord are connected to relay nuclei, which are connected to the cerebellum and to cortical projection areas, which are connected to secondary sensory and to higher association areas and basal ganglia nuclei at the 'top'.

This hierarchy is not strict – there are shortcuts going up or down, and inside a nucleus or a brain area there are receiving and sending neurons, and neurons connected locally, which are further removed from the periphery than others. For our purposes it gives a good rough approximation, though.

3.2. Information processing is organized hierarchically

In general, lower levels of the sensory hierarchy represent concrete, low-level physical features that can be extracted from the stimulus (bottom). Lower levels of the motor hierarchy represent movements of individual muscles (bottom). Higher levels in the sensory hierarchy represent more abstract category information, or aspects of the stimulus situation that are not directly observable, typically at a larger spatial and temporal scale as well (top).

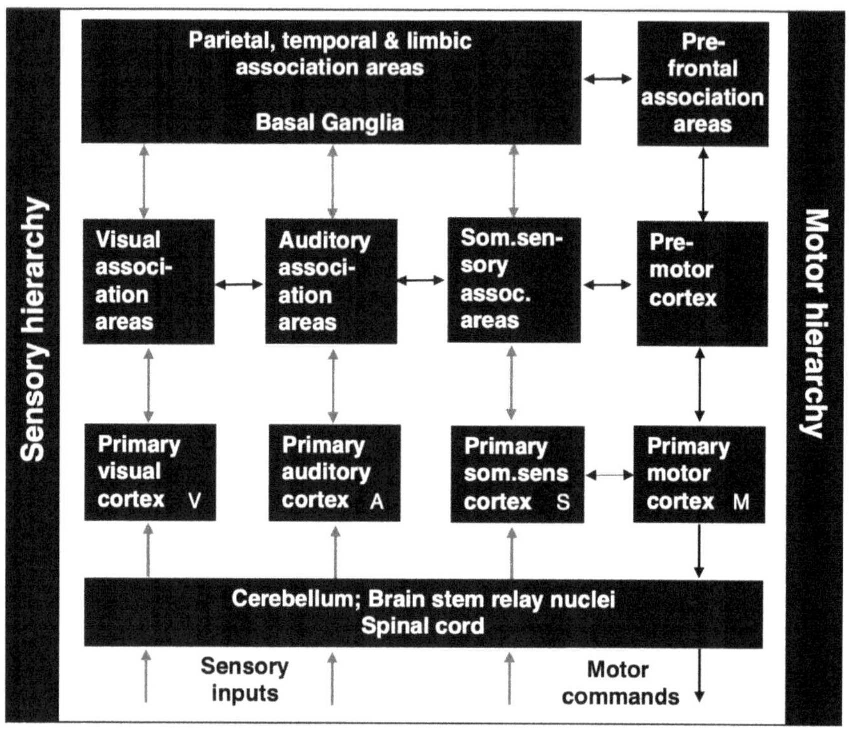

Sensory hierarchy
Parietal, temporal & limbic association areas
Basal Ganglia
Pre-frontal association areas
Visual association areas
Auditory association areas
Som.sensory assoc. areas
Pre-motor cortex
Primary visual cortex V
Primary auditory cortex A
Primary som.sens cortex S
Primary motor cortex M
Cerebellum; Brain stem relay nuclei
Spinal cord
Sensory inputs
Motor commands
Motor hierarchy

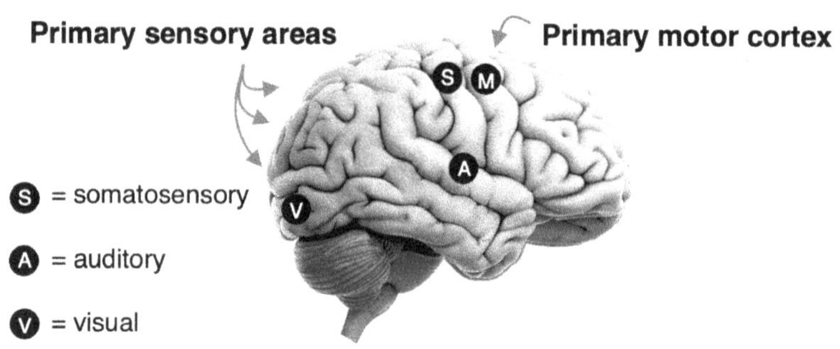

Primary sensory areas
Primary motor cortex
S = somatosensory
A = auditory
V = visual

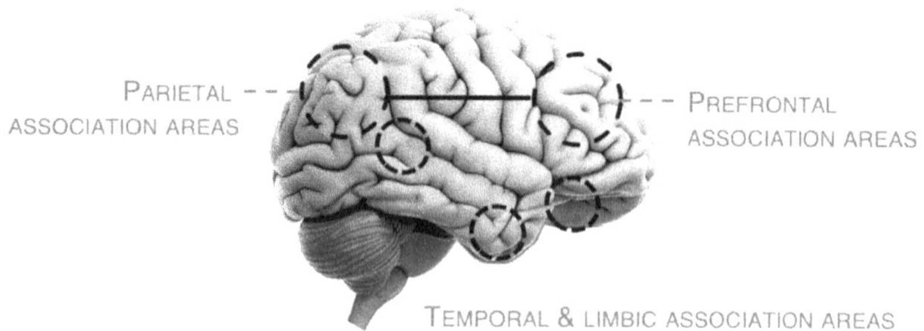

FRONTOPARIETAL ATTENTION NETWORKS
PARIETAL ASSOCIATION AREAS
PREFRONTAL ASSOCIATION AREAS
TEMPORAL & LIMBIC ASSOCIATION AREAS

Figure 7-1. Schematic outline of the neuroanatomical hierarchy. Note the bidirectional and lateral connections in the hierarchies.

Primary sensory cortical areas (visual, somatosensory, auditory...) are the waystation for signals from the sensorimotor periphery and relay nuclei deeper to the cortex.

Primary motor cortex is the final waystation of motor programs, sending motor commands to the brain stem and spinal cord.

Multisensory integration, higher-level motor planning and memory is dependent on association areas.[129]

Higher levels of the motor hierarchy represent extended movement sequences and the hierarchical dependencies and (re)arrangements of movements, and more abstract action goals (top). Typically, the higher-order motor sequences and hierarchies span more time and space as well.

4. Wired for prediction

This piece of the puzzle of how the brain works is one of the deep unifying themes in current neuroscience: predictive processing. It is a hot topic because it may give a common vocabulary for cognitive and computational neuroscientists, neuroanatomists, experimental psychologists, as well as engineers working in machine vision, machine learning and computer science. Prediction could therefore become a unifying concept to understand cognition and brain function in complex tasks. How is predictive processing visible in brain design?

4.1. Bidirectional connectivity

The brain is not a one-way street. Anatomically, whenever one area connects to another, there are generally recurrent connections back to that area. Unidirectional connections, where brain area A sends out nerve fibers that synapse in area B but area B does not connect back to area A, is the exception, not the rule.

(The cells sending back axons from B to A are not necessarily axons of the cells receiving the synaptic input, though. The input cells and output cells of area B may be connected via interneurons modifying the signal so that the output neurons of B are reporting back to A the results of some operation on the signal from A, combined with other sources.)[130]

4.2. Predictive coding

Consider the signals arriving from sensory organs to various relay nuclei in the brain. How do they relate to the signals that get transmitted from the waystations to higher up?

What these lower levels actually report higher up, the bottom-up feed sent deeper into the brain, is not a complete copy of the sensory features they have registered and analyzed. This would be ineffective, as the central systems higher up in the hierarchy would be totally swamped in converging data. The lower levels need to send only 'need to know' information, not everything they are seeing. But how do they know what the higher level needs to know?

Try to imagine what it would be like hearing every rustle of leaves, every breath of wind, every piece of conversation and *at the same time* seeing every blade of grass, every reflection of the sun from every surface, every fleeting expression on every face, and *at the same time* feeling the pressure of the ground on the soles of your feet, the sun in your face, the feel of your clothes...

Somehow, somewhere, parts of your brain are registering all this, but you are not aware of it (unless, say, there is an ant crawling in your shirt – suddenly you do become aware of this unexpected feel of your clothes...).

What we are aware of at any time is some highly filtered, schematized, abstract sense of the situation as a whole. How does the brain decide *which* parts of the picture to filter out, and which parts to report up the hierarchy, so as to build the situational awareness of only the relevant stuff?

Part of the story is the clever way that our brain is designed to save on channel capacity: predictive coding in the sensory systems.

As a toy example, let's say the visual cortex is analyzing the contours and colours of an object, and is able on its own to determine it is a shoe. While this would allow the brain to spare on forward connections, it would *also* mean that the visual cortex already would have to 'understand' the category structure of *everything in the world*, including shoes.

And because you can also recognize a shoe by its feel, the concept of a shoe would have to be understood by the somatosensory system as well… and the auditory system (judging by the sounds of the steps, is she walking barefoot or with shoes on?)…

But what happens, instead, is that only the higher centres of the brain really 'understand about shoes' as such. But based on the knowledge about shoes they are able to predict some lower-level (visual, touch, sound) consequences of the presence of a shoe vs. absence of a shoe. What the lower level does, then, is to compare this prediction with the actual perceptual input they are getting, and then report forward only the prediction error (if any).

In the hierarchy, the somatosensory lower level handles, for example, details of the feel of the shoe. And this is itself at a level higher than registering raw sensory feedback from touch and pressure receptors: it is itself already predicting this more basic incoming raw feedback signal, based on its own internal models.

The idea here is that the systems lower in the hierarchy are not inferring, abstracting or deducing some new properties from the *bottom-up* information, and then reporting this new more complex information forward. Instead, the higher levels make guesses and send down information that tell the lower levels (help them predict) what features they should be seeing. The lower level then compares this prediction to what it is actually seeing (feedback), and reports back the discrepancy (prediction error). This process is predictive coding.

This organization of perception and memory into a predictive coding hierarchy has many useful consequences. First, the lower levels only need to trade in the specific aspects of the world they understand, while the higher level does not need to trade in all the specifics. The eye gets to see the visual form of the racetrack whizzing by, and the hand gets to feel the haptic torque of the steering wheel. Visuomotor integration into the concept of 'an oversteer moment' is being done only at the higher levels, but that higher level can leave out some of the variability in the details to be handled at the bottom.

Second, if the predictions are accurate (the predictive internal models higher-up are up-to-date and able to correctly predict lower-level features of the situation), then the lower levels, experiencing no prediction error, send forward nothing. This saves channel capacity and energy, as the higher levels are only sent information on a need-to-know basis. Only 'newsworthy' signals get propagated further. You really start to notice the oversteer moment only as it starts to get out of hand. Before that it's just a dab of opposite lock. Business as usual for the racer's brain.

This also gives the brain a very natural way to determine which higher-level concepts (chunks) best match a situation: it's the one(s) that best predict lower-level features – in other words receive the least prediction error. So, in the race for memory trace activation and consolidation into long-term memory, the activity patterns that produce the least prediction error should be the active ones, and get stored.

(The word 'race' here is not a pun, by the way – many cognitive models are literally called race models because the representations compete for activation in a winner-take-all manner, such that the representation that reaches some threshold activation first prevails, and the others get inhibited.)

5. Learning from prediction error

The predictive processing view also gives a unified account of learning. What learning *is*, on this view, is synaptic modification (establishing long-term memory traces) where internal models are updated to improve the prediction of lower-level features: the feedback signal for learning, at every level, is prediction error from the level below. The goal of learning (all learning, at every level) is prediction error minimization in the future.[131]

These five 'laws' are the most condensed form I (OL) can give to the question 'How does the brain work?' They are not the whole story (obviously), but they give you a fairly good idea of how complex brain functions are organized: in some way most brain structures actually do more or less the same thing.

In order to be able to be both fast and accurate, the operation a brain system performs just has to be mostly a memory-retrieval operation (or at most a very, very simple inference), because memory is fast and reliable, but thinking is slow and unreliable. Any process that operates fast and reliably, such as perception, has to be mostly memory, not thinking.

Moreover, if the brain is wired as a hierarchical prediction machine, most of this is going to be comparing retrieved information to incoming input – i.e. prediction error computation. Based on the information embodied in the local synaptic connections (memory, or internal models, which are all different names for the same thing, really), a system takes predictions from higher areas, and feedback from lower areas, and compares them. It then reports prediction error (only) back up, and issues its own predictions to the level below.

So that's it then? Well, not quite. It's a bit like saying that every mechanical part of an engine pushes around and gets pushed around by other mechanical parts. True on a general level, but not specific enough. The brain consists of hundreds or even thousands of distinct components (with weird names in Latin and Greek). And each specific structure in the brain – different patches of cortex, different subcortical nuclei, the circuits they form through neural tracts… – do contain special types of neurons and are very selectively connected to the rest of the brain (the brain is sparsely, not fully connected).

This suggests each structure might have very specific roles in cognitive function. The structures and their connections are different because they have different cognitive roles. Local systems will differ in terms of what information is being retrieved from memory, and what inputs they are getting compared with.

FURTHER THOUGHTS ON LOCALIZATION

In subjective experience, the mind appears to consist of different faculties: perception, imagination, feelings, memory, the ability to plan ahead and reason, and to produce voluntary movements… At multiple time scales, from split-second decisions to memories lasting a lifetime, it's these things that make us who we are.

The brain also consists of distinct parts. Neuroanatomists divide it into the brain stem, cerebellum, subcortical nuclei, various areas of the cerebral cortex… and each of these comes with innumerable subdivisions. At multiple spatial scales, from individual neurons and local circuits to larger systems of billions of cells, these structures also make you who you are.

A natural question arises as to the relation between faculties of the mind and the components of the brain. How are they aligned? This is the question of the *localization* of mental functions. Is there perhaps a special component for every function – visual 'centres' and memory 'stores'?

While it is tempting to think in these localizationist terms, the complex systems view of brain function teaches you otherwise. And while we *do* want to know '*how the elementary ingredients of the* [mind] *correspond to the elementary functions of the* [brain]',[132] what the elementary functions of the mind *are*, in the first place, is not something that is directly given by experience.

Remember: in conscious awareness we experience only the outcome of a tremendously complex process – countless elementary functions working away in the background. And now you know this background consists of a hierarchy of predictive machines. So, you are trying to localize hierarchical predictions of things like pressure on the skin or patterns in the visual field (in the racecar analogy think slip angle, think brake fluid pressure, think fuel vaporization), and not subjective experiences of things like feeling the point of balance (don't think good running, don't think steering and handling).

'What does brain structure X do?' It is such a straightforward and natural question, for both neuroscientists and laypeople. What does the visual cortex do? Or, conversely, 'what brain structures perform psychological function Y?' Is it, perhaps, the visual cortex where vision happens?

But while the question of localization is easy enough to pose, it is not so straightforward to answer in a way that gets it right. Not just because the brain is very complex and contains many, many interacting structures – and it does – but because the logic of the answer requires you to think of the brain as a complex functional system. And this is something people are not used to.

First, what most people will probaby think about when they hear the words 'local brain structure' is a piece of the brain bounded inside a small volume. Something you could scoop out with a spoon. But most processes neuroscientists are interested in are actually 'localized' into highly *distributed* networks. Threaded, gossamer-like systems that can span large distances. So the picture is different right there. The visual cortex (which you could scoop) and the visual system (which is gossamer-like) are not the same thing.

This oxymoronic-sounding 'localization into distributed systems' does not mean the process is a diffuse mass action, though. The function Y (vision) can still be localized in X (visual system). Like, the function of braking (Y) can be localized in the circuitry (X) spanning the brake pedal, ducts, master cylinders and calipers etc. The brake system just happens to span large distances and many components. That does not mean braking would be diffusely distributed equally in *every* part of the car.

This analogy should also encourage you to look at every part of the brain in terms of its connectivity – which parts it communicates and interacts with, and how. And this is the right way to look at it. Because it then leads you to ask: 'what does the brain structure X do in the complex system it is embedded in?' This is a more sophisticated question. (Like: 'What is the role of the brake master cylinder in distributing braking effort across the front and rear axles?')

Let's take the analogy a little bit further to bring the point home. You can look at a carburettor, say, and describe its 'local anatomy' all you like... But it's role in the complex function of producing power in terms of gas flow, fluid vaporization, air-fuel ratios and so on – *this* is what you need to know if you want to figure out what the carburettor does.

If you want to figure out how to make your car go fast(er), there are fundamental design choices for the engineer. And the engineer can consider modifying the carburettor, and look at its design from physical first principles. But any garage-bound engine tuner will also understand, and will need to understand, how the carburettor works in the context of the system. For example, if you are able to get more air into the engine you also need to get in more fuel – neither on their own will help much, because it's their balance, the composition of the fuel-air mixture, that counts as much, or more, for proper combustion as the volume of material.

And here's the thing: the design of the carburettor *only makes sense* when you look at it in the context of fuel-air mixing, cylinders and pistons, ignition timing, powerline behaviour etc. And it is exactly the same thing with the brain. Not just similar, the same thing. The carburettor is not here a metaphor: a brain, like a car engine, are both literally examples of complex functional systems. Meaning roughly: there are many parts, and the effect of changing the behaviour of any one part critically depends on that part's interactions with other parts.

This analogy should help you see how complex information processing or mental phenomena like 'emotion' or 'driving skill' is not localized in brain centres any more than 'horsepower' is localized in the carburettor. The carburettor for sure makes a difference to horsepower (while, say, damper valves or wing angles don't), and the contribution is specific (different from the contribution of the spark plug)... but it only makes this contribution in the context of its connectivity. Exactly the same idea applies to 'localization' of a cognitive process or other mental phenomena in brain activation: you just cannot localize a complex cognitive function into a 'brain centre'. You need to look at the thing as a system.

The second thing that is hard for people to get right is what processes you are trying to localize in the first place are. 'Braking', not 'making it stop'; 'air & fuel flow', not 'making power' or 'making it go'. If you are, to begin with, very confused about the nature of the function, then knowing where it happens – localization – is not even possible, because you are posing a confused question.

It's like asking 'where exactly, in which component or system, in this classic 1960s Ferrari racing car is its collector value localized?' It's not that the component is unknown, or the system too complex to understand – the question itself does not make sense. The components of the car do not play a role in 'generating value' in the same sense that they play roles in generating power.[133]

'BRAIN ACTIVATION' IS NOT BRAIN ACTIVITY!

And one final thing. Before we take a first look at functional brain imaging research on 'localized brain activation', it's really important to be clear on the difference between, on the one hand, what brain areas become *activated* in different tasks (this is what brain imaging results can tell you) and, on the other hand, which parts of the brain are *active* during the task (which is not something brain imaging results tell you, at least never directly on their own; yeah, it sounds weird).

If you look at the fine print, brain science results are hardly ever about brain *function* as such, only brain *activation*. So, what is brain 'activation', really? To answer this, we need to look more deeply into how brain scientists go about discovering facts about 'local brain activation'.

The interpretation of brain imaging results requires that you become comfortable with the idea that results about brain activation are always comparisons between two things. Scientific measurements always are, really – it is just not so obvious in every case. But think about it: you *compare* the distance between your head and toe to the distance between the ends of a ruler, and this gives you your length; you *compare* the time a racecar takes to cover a quarter mile to the time your stopwatch takes to tick one second, and this gives you the elapsed time, etc.

It's the same with 'localized activation' that brain scientists use to pin down specific brain structures involved in specific psychological processes: you are *comparing* brain function in different situations or different individuals... and the *difference* is the activation shown by brightly coloured pixels in brain imaging pictures. (The actual techniques for doing this for complex time-series data require some advanced statistics, but the underlying logic is not at all mathematical.)

A person is shown a disgusting picture of insect-infested food, say, and blood flow in a certain part of the brain, the insular cortex[134], measurably increases. Show them a horrific accident scene and the amygdala lights up. The emotion of experiencing disgust has now been localized into the insular cortex, and fright in the amygdala, right? Wrong. It's like localizing turning to the steering wheel and stopping to the brake pedal. The insula 'lights up' for disgust, but pictures of this activation do not show the whole circuitry for disgust, the 'disgust system' (if there is such a system). The brake light lights up at the braking point, but, you know... it's not the whole system.

Let's work through an example that shows how the logic of brain activation works on the basis of *contrasts.* Contrasts are comparisons. Between one set of measurements of brain activity (from a specific task, or a particular group of participants) and another set of measurements of brain activity at the same locations (in another task, or a comparison group of participants, acting as 'statistical control'; **Figure 7-2**, overleaf).

Suppose a brain researcher is interested in the phenomenon of mental imagery. Let's say her particular interest is in mental rotation: being able to voluntarily, on demand, make whatever you are imagining in your mind's eye rotate around some axes so that you get a 'mind's eye view' of it from different directions. (It is likely that at least some of the same processes are involved when the racing driver runs his 'mental movie' of a racetrack, so it's more relevant to the racer's brain than disgust.)

So, she will proceed to get some participants into a lab with an fMRI brain scanner and give them specific instructions on how she wants them to carry out 'a mental rotation task' whilst lying inside the device. For example, she might ask the participant to conjure up in

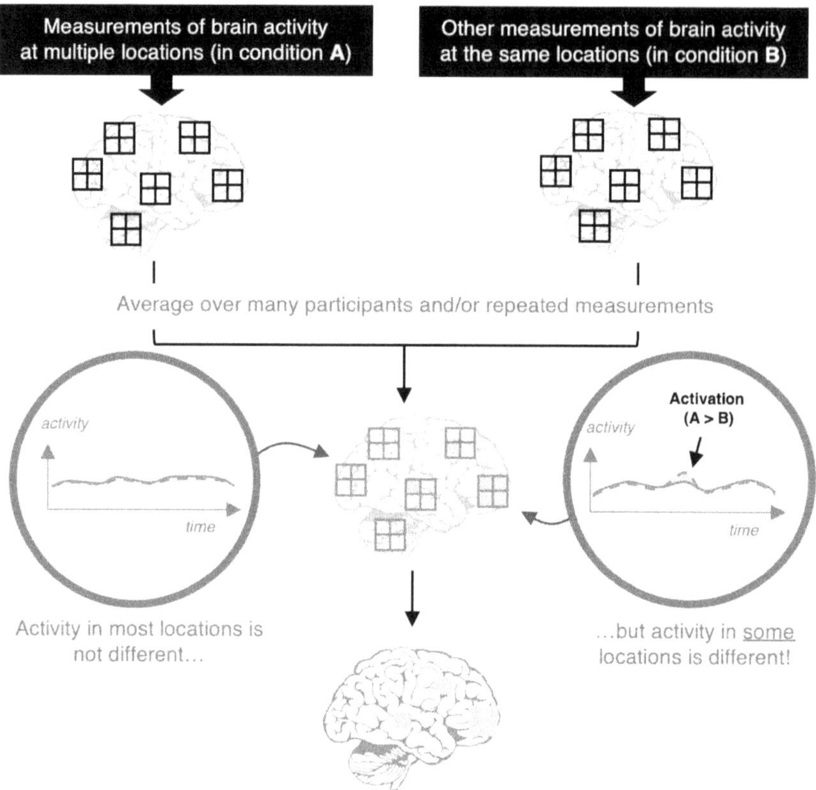

Figure 7-2. The logic of the method of contrasts. More complicated and fancy ways of analyzing patterns of brain activity across the brain have been developed, and are now becoming the standard way to define activation in terms of distributed networks, not local foci. But the underlying logic, statistically contrasting blood flow measurements in one task or one set of participants (A) to blood flow measurements in another set (B), is the same.

their mind's eye a square, and then make it rotate clockwise and counterclockwise when presented with a sound cue. (This is a simplified example for illustrating the general idea – in actual practice a lot of thinking and expertise go into designing the task instruction.)

As the participant is doing the rotating, the scanner measures blood flow in the different blood vessels all across the brain. The idea is that more blood flow is directed to regions where brain structures are working harder.

After many measurements from a bunch of participants have been collected, the scientist statistically analyzes the data and produces images of *loci of activation* ('loci' is the plural of 'locus', meaning place).

The first thing to notice is that what she is measuring is not neuronal signalling in the structures of interest per se; that is, she is not measuring the processes transmitting and operating on the information (the action potentials and synaptic transmission between neurons). What she is measuring is blood flow. (Something more like blushing of different parts of the brain as blood moves to areas of higher activity. Well. *Strictly*, she's looking at the oxygen content of the blood, as neurons use up glucose; you can find out about the details in a neuroscience textbook.)

The blood flow signal is nevertheless generally considered a good indirect index of synaptic transmission, the most energy-sapping bit of brain activity. A bit like a race engineer might use speed and RPM signals as a proxy for gear selection.

The second and more important thing is that if she observes 'activation' in some part of the brain it does not imply that the rest of the brain is not doing anything! Only inside a dead brain (or a gangrene in a living brain) is there no brain activity, and no blood circulation. The activation is really a signal *difference* (on the order of +/- 1%) as the local blood flow in a small volume of brain tissue is compared across different times, and between different tasks. Neuroscientists call the small volumes voxels, analogous to pixels in a flat 2D image. A voxel with 1% more blood flow during mental rotation, when compared to the flow level in a control task, will 'light up'.

And the really important part here is that the comparison is not to a dead brain, with no brain activity. Which voxels show activation in task A depends on the chosen comparison, not just A itself; because it's the *difference* from the 'statistical control task' B measurements that you are seeing. (Often you see the contrast indicated by A > B, read: 'A greater than B'.)

The scientist has to decide what task condition is to be contrasted against the task condition of interest, to get at the desired difference.

Knowledge of this decision is what you should then use to interpret what A > B activation[135] means.

Suppose the task of interest A requires mental operations x, y and z. These could be for example conjuring up a mental image and rotating it. Then suppose the control task B requires operations x, y and q (conjuring up a mental image and holding it steady and clear in the mind's eye). In this case the comparison of A > B will show activation in some areas or structures relating specifically to the mental difference z vs. q (rotating vs. steady). It will not show any activation in structures performing the shared operations x and y (visualizing the image in the first place). This means that you will not see all the brain activity involved in performing task A.

But this is actually a *good* thing, because the scientist can now be more precise: rather than saying that there are all sorts of processes all around the brain that 'relate to A', they can say that 'activation' of z (and/or deactivation of q) relates specifically to the operation of rotating the image. In this way, by analyzing how A and B are different at the cognitive level, and *only* by analyzing how A and B are different at the cognitive level, can one begin to try to pinpoint the neural substrates of elementary operations that can be localized.

The snatch is that it can be misleading if one then just thinks 'hey the brain was scanned while doing task A, so the activation they see is an index of the brain doing A'. And one would be completely wrong in thinking that *only* the activated brain region is single-handedly responsible for the ability to perform task A!

All we are seeing is that the comparison of measurements in task A vs. task B shows some areas to be activated by A when compared specifically against B. (Or, as we'll see a bit later, how some areas are activated in an expert racing driver group, when compared against neurotypical controls.)

To be able to say what that particular area does, though, you additionally need to have a precise analysis of A, and specifically *how exactly* task A is different from task B. (Or what exactly the expert is doing that the control subject is not doing.) Only this tells you what elementary ingredients of the mind A and B differ in.

Finally, let us note that a cognitive analysis of A into those more elementary processes, x y and z will refer to operations that are likely to be not immediately obvious, are perhaps unintuitive, and will be difficult to put into words. (This is why cognitive modelling, like any other field of science, relies on mathematics and computation as the language to express hypotheses; to define what operations, in our example 'rotation of an image', fundamentally consist of.)

Some consequences of all of the above bear pointing out, because they can run counter to the intuitions reinforced by the popular-science view of 'brain activation'. Getting these sorted is fundamental:

i. You cannot localize a complex mental function into a brain centre for that function. A complex skill like driving ability does not reside in a dedicated 'brain centre'.
ii. Only 'elementary ingredients' can be localized, and it is not always obvious what the elementary ingredients are. This is why it is so important to analyze complex behaviour into its cognitive constituents, like we have been doing in **Chapters 1 to 5.**
iii. Observations of localized brain activation are not the same as the organization of the underlying brain activity. This is not just a play of words – 'brain activation' vs. 'brain activity'. (In the technical brain imaging sense, activation simply is not the same thing as brain activity or brain function, as the latter are usually understood.)

It is unfortunate that the words 'activation' and 'activity' are so alike. In fact, the confusion is so pervasive that the brain scientist Bradley Postle in a textbook on cognitive neuroscience has suggested that it would be more apt to call brain imaging patterns 'blobs', not 'activations'!

Now, regarding these blobs, there is one more subtle point we should mention, which is about what counts as 'local'. Most of the blobs you see are typically produced by what's technically called univariate analysis methods. This means that you are comparing the signal coming from one local brain structure (or one voxel) in the different tasks (A vs. B). Basically, it means you can draw a rectangular box around the activation pattern (as you can, say, around the steering wheel, or a brake caliper; you could scoop it out).

More sophisticated statistical methods allow brain scientists to look for more complex 'multivoxel' patterns: not just whether local blobs are activated or not, but whether they are *co-activated* across subjects, and then looking for *similarities* in co-activation, not just differences in activation. This can reveal more distributed networks of co-activation.

The main difference is that multivoxel patterns are networks of multiple linked components (like the steering system or the brake system; brain scientists call them functional networks[136]). The shape of the activated region(s) does not fundamentally change the reasoning, however. The underlying logic of contrasts is similar.

From a cognitive science point of view, the crucial issue is not whether you are mapping a complex psychological construct onto a blob or onto a network – the crucial issue is whether you are analyzing the complex psychological construct into the right elementary cognitive operations. And knowing what brain area or circuit is 'activated' does not tell you this. Which brings us to the final point:

iv. Having a methodology for observing localized brain activation is not the same thing as having a theory of the constituent cognitive processes.

And you do need a theory of cognitive processes to interpret the localized brain activity – whether that activity is a blob or whether it is a network. This is why we spent the whole of **Parts I-II** of the book analyzing the racer's complex skill and cognitive processes, rather than delving straight into the racer's brain!

BRAIN ACTIVATION IN DRIVING

Keeping all these caveats over 'brain activation' in mind, what can be learned from work on brain activation in driving? There are to date a few dozen brain imaging experiments that have looked specifically at driving (mostly everyday driving scenarios, in simulated environments, for obvious practical reasons).[137] A statistical meta-analysis of this work[138] has identified brain areas commonly reported as activated by driving (or just watching in-car footage), when compared against various control tasks.[139]

These include parts of the cerebellum, certain visual areas (involved in higher-level analysis, such as visual motion), motor areas of the frontal lobe, and parts of the so-called salience network (anterior insula and posterior cingulate).

The first thing to note is that there is consistency between experiments on which brain areas are commonly activated in driving tasks.

(Remember, one must bear in mind that there are differences in what 'statistical control' tasks were being used to compare 'driving' to.)

The authors of this meta-analysis classify additional areas seen in some contrasts, but not others, into a hierarchy of operational, tactical, and strategic levels [140]. These align closely (but not exactly) with control, guidance and navigation, and derive from influential traffic psychology theories of the 1980s.

The strategic level is route selection and goal setting at a scale of minutes or more, for example weighing expediency vs. driving pleasure in whether to continue on the motorway or to take the back roads on the way home. The tactical or manoeuvering level refers to decisions – often habitual ones – on a time scale of a few seconds, e.g. whether to overtake. The operational level is the physical execution of the decisions made at the higher levels, e.g. turning on the indicator or changing gears when slowing down to take the off-ramp… to do the back roads… obviously.

Within this framework, they associated the strategic level to (get ready for some Latin here) the ventrolateral prefrontal cortex, and parts of the so-called ventral attention network (temporoparietal junction and superior temporal sulcus). They associated the tactical level with premotor areas of the frontal lobe (considered to be relevant for movement sequences) as well as certain visual areas (related to both visual scene analysis and visuomotor control). The operational level they only relate to activation of the visual cortex (and certain subcortical structures), though here it must be considered that many of the 'driving' tasks in these experiments are in fact viewing tasks – not active motor control tasks. (Partly due to practical limitations of brain scanning.)

One study to date[141] has looked specifically at brain activation and brain anatomy of expert racing drivers and compared them to neurotypical controls. It used a slightly different imaging method than most studies, called inter-subject correlation (ISC compares local signal change across time and between subjects, to see if one group of subjects has specific parts of their brain more 'in sync' with one another than the other group).[142]

Professional racing drivers and control participants watched in-car footage of a Formula One car on F1 tracks and were instructed to 'imagine themselves driving the racing car'. The ISC analyses reveal activation in terms of which parts of the brain show blood flow changes that are more synchronized to the on-screen events (in either group) (**Figure 7-3**).

In both groups, activation (synchronization across the participants) was observed again in visual areas, precentral cortex, posterior parietal cortex and also parahippocampal cortex (containing a 'place area' that is commonly found to activate in many types of scene-viewing tasks).

In the racers only, however, there was *also* activation in fronto-parietal areas and the dorsal visual stream, including the prefrontal cortex: the tip of the prefrontal cortex called frontal pole and the ventrolateral prefrontal cortex that is usually called Broca's area. Also important was activation of the retrosplenial cortex.

Wow… now that's a mouthful. So, what does all this Latin *mean?* To develop meaningful interpretations (that is: cognitive level hypotheses), you need to think about the differences between how a racing driver sees the world vs. how a normal driver sees the world. Oh, this is exactly what we have been doing so far. How lucky is that?

When a racing driver views in–car footage, she is not merely mentally simulating 'steering in the direction of the bend'. It's a much more complex process, so her brain will engage in the task in a quali-

Figure 7-3. Bernadi et al. (2014) investigated the differences between brain activation in racing drivers, when contrasted against brain activation of normal controls, whilst watching on-board footage of an F1 car. There were many brain structures that showed activation in racing drivers but not in the control subjects. Many of them are known to be involved in visual scene analysis, focusing of attention, orienting in space, and action planning. (Note: activation here means that blood flow was more synchronized to the events on the screen; this figure only shows within-group activations, not the racer > control subject contrast).

LH, left hemisphere; RH, right hemisphere; aPFC, anterior prefrontal cortex; dlPFC, dorsolateral prefrontal cortex; vlPFC, ventrolateral prefrontal cortex; IPL, inferior parietal lobule; SPL, superior parietal lobule; MTG, middle temporal gyrus; POC/PPC, parieto-occipital/posterior parietal cortex; PHC, parahippocampal cortex; PCC/RSC, posterior cingulate/ retro-splenial cortex; V1+, occipital visual areas.

Brain imaging figures from Bernardi et al. (2014) (CC-BY-40)

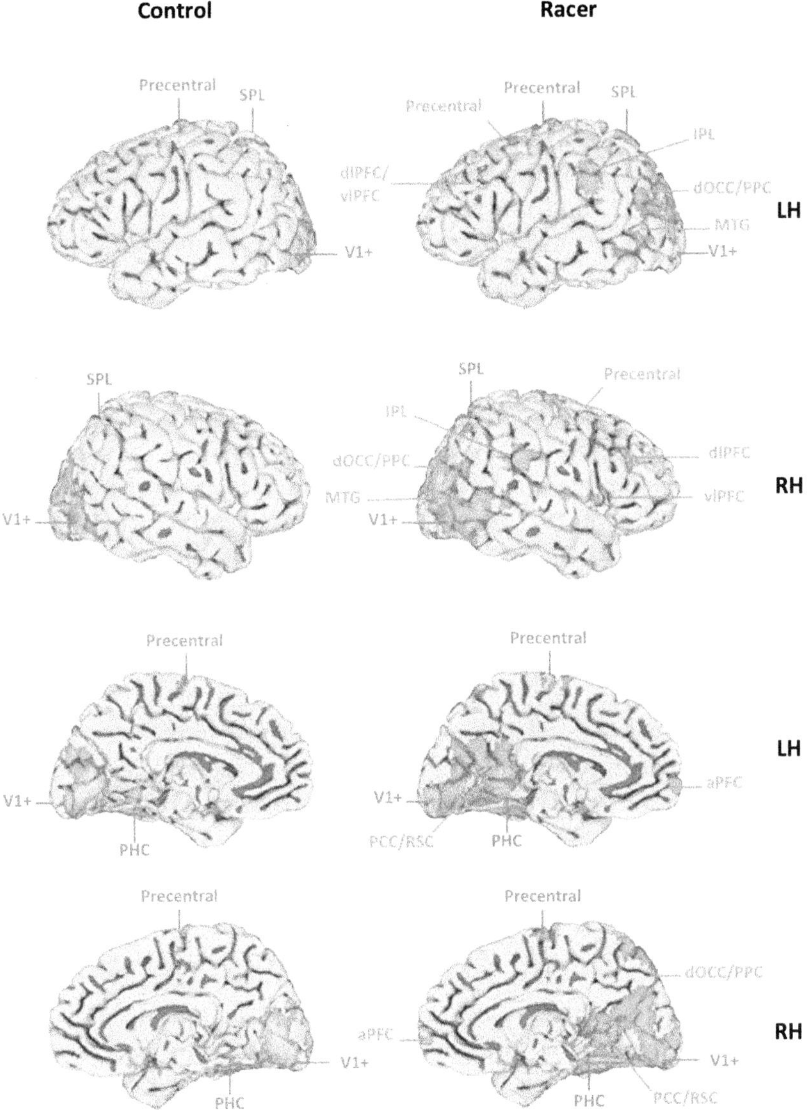
Control
Racer
Precentral
SPL
V1+
Precentral
Precentral
SPL
IPL
dlPFC/
vlPFC
dOCC/PPC
MTG
V1+
LH
SPL
V1+
SPL
Precentral
IPL
dOCC/PPC
MTG
V1+
dlPFC
vlPFC
RH
Precentral
V1+
PHC
Precentral
V1+
PCC/RSC
PHC
aPFC
LH
Precentral
V1+
PHC
Precentral
dOCC/PPC
aPFC
V1+
PHC
PCC/RSC
RH

tatively different way. (Remember our discussion on how a casual viewer and a pro may *look* at the same TV broadcast but *see* completely different things?)

The frontoparietal activation was in the dorsal attention network – its exact location does not matter here – and may reflect higher, more focused task engagement in the racing drivers. When it is considered in the context of synchrony being observed in Broca's area (lateral prefrontal cortex) and the retrosplenial cortex, the picture becomes particularly interesting, because of the kinds of mental operations these structures are commonly taken to support.

Broca's area is generally thought to be involved in processing hierarchical organization e.g. in language and music. But note that hierarchical organization is also a hallmark in sub-goal setting. That is, chunking!

The retrosplenial cortex, on the other hand, is part of the brain's scene representation and navigation networks. It sits in the middle of the rear half the brain – though again that's not so important – and is thought to be involved in piecing together location knowledge from multiple snapshot views of the scene.

That is, it's needed to create an understanding of places, when different views of the same locations and objects are seen from many points of vantage. It's also involved in spatial memory retrieval needed to orient oneself within a large-scale space too big and complicated to see from one point of vantage (like a racetrack!). Especially so when you need to orient yourself on the basis of familiar landmark knowledge (like reference points!).

So, could it be that the task synchronization in these areas is actually giving us the first direct window into chunking processes that the racer's brain uses to 'read' the track layout?

And there's more: this study also explored *structural* differences in gray matter size between the racers and the control group. Local gray matter density differed in a number of brain areas, including the lateral prefrontal cortex and the retrosplenial cortex. Particularly impressive was that in the racing drivers' gray matter density in the left retrosplenial cortex correlated with a 'driving proficiency index' (percentage of races finished on the podium).

The cross-sectional design used does not allow one to directly conclude to what extent the observed differences may have been produced by the race driving experience of the expert subjects ('prac–

tice effects': brain areas growing in response to training), and to what extent they may reflect pre-existing skills and adaptations that have enabled those individuals to pursue a career in racing ('talent': those individuals who are already wired differently from the outset getting the breaks, and over time becoming more proficient).

Studies in musicians has shown that they have enlarged motor cortices (the motor cortex was another area where the racers showed more gray matter volume than the controls), and that the size of the cortex actually does change in response to sustained practice. (And in a specific way: for a violinist it's the motor cortex responsible for controlling fingers of the left hand!) [143]

Also, seminal work on expert taxi drivers, London cabbies, have shown something similar. This job requires remembering the layout of a super-complex network of streets (London is definitely not arranged in a logical and principled manner like, say, Barcelona or even Paris.) They also need to calculate alternative routes between arbitrary locations, and re-compute routes on-the-fly, for example when there are road-blockages or traffic jams.

These people are true navigation/strategic-level experts, trained with a gruelling regime called The Knowledge. They are required to know every street, street number, hospital, hotel, club, tourist landmark, cinema, school etc. within six miles of Charing Cross. And more than that, when taken to a particular location they must be able to plot an efficient route to any given destination.

And so, their hippocampus – a temporal lobe limbic structure involved in spatial localization and the encoding and retrieving episodic memory information – shows to be larger in size. And the size is correlated with years on the job, and also responds to training by growing in size! (One cannot know positively to what extent the constant training would cause everyone's hippocampi to grow like that, or whether 'the ones who just don't have the hippocampus for it' are not able to stay on the job.)

Actually, it is even more specific than that: it is the *navigational* part of the hippocampus that is enlarged: the rear end of the hippocampus, connected – via the retrosplenial cortex – to the dorsal attention network. (fMRI activation analyses showed that this structure was indeed activated in the cabbies in route-planning situations.)

In contrast, the concept-memory-supporting anterior parts of the hippocampus connected to the temporal lobe are actually reduced in size. And the expert-navigator cabbies in fact do perform worse-than average on concept-memory tasks.

This is an example of the *domain specificity* of expertise, and the very specific contributions of local brain structures to specific functional circuits. As with muscles, practice develops specifically those parts of the brain (and body) that are needed for a specific job. There is no such thing as 'general training' for cognitive ability – any more so than for 'physical ability'. Which is to say: there may be some, but that is not what makes you *really, really good* at anything.

LOCALIZATION IS NOT A THEORY

This point is important, so at the risk of repeating ourselves, let us repeat. In spite of the remarkable specificity of, say, the role of the (posterior) hippocampus in route planning, it's not the case that (posterior) hippocampus on its own would be doing route planning for the taxi drivers. Likewise, you should not infer that the prefrontal cortex or retrosplenial cortex on their own would be doing chunking of a scene into reference points for the racing drivers. The fact that these localized structures are activated (in specific contrasts) does not mean that the entire complex cognitive function can be localized in these specific brain structures.

And more than that, *localizing* a function is not the same as *understanding* that function. Scientists are just beginning to figure out *what* computations the posterior hippocampus or the retrosplenial cortex are actually performing – their unique contributions to wayfinding behavior.

Where a brain structure is located is well known, and the same for every individual. *What* elementary operations can be localized there is often not well understood, and what complex skills these operations support is not the same for all individuals.

Okay. You've made it this far. Now is the time to see how it all fits together.

Chapter Eight

Designed for Speed – the Racer's Brain

THE RACING CAR CAN BE CONSIDERED a technological extension of the human body. This goes for any type of vehicle: cars, bicycles, skateboards, skis, roller skates... They all modify movement capabilities, and the feedback you get from motion. Even shoes separate contact forces from control forces (only when walking barefoot are they the same). And as we will see in a moment, just like your shoes and clothes, the racing car may be quite literally treated by your brain as an extension of the body; you 'wear it' in quite a literal sense.

The design of the human brain enables people to develop expertise in very many forms of high-speed locomotion. How can this be? After all, the difference between what happens at the contact points – steering wheel, pedals – and what happens at the control surfaces – the contact patches – can be quite dramatic.

At the mere extension of the ankle, a little pressure at the sole of the foot, somewhere behind you the car is putting huge forces on the ground, and hurtling you to speeds that would have been almost unimaginable to ancestors just a few generations back... let alone ancestors in our deep evolutionary past, where many of our basic brain mechanisms hail from. Amazingly, the core processes responsible for the brain's capacity for guiding movement in 3D space are well up for it.

And make no mistake, despite its apparent simplicity, this is a formidable capacity, the flexibility humans display in so quickly and so accurately selecting a path through a complex three-dimensional scene, and coming up with the right motor commands to take that path is the envy of roboticists and AI researchers.

This ability is not, of course, limited to the job of the racing driver: it is evidenced in downhill skiing, mountain biking etc. And you can see it in other species too: the cheetah racing the antelope, the fox giving chase to the rabbit in the undergrowth, the buzzard flying through trees in pursuit of the starling…

This suggests that animal brains have developed, over hundreds of millions of years of R & D called natural selection, fairly general and yet highly efficient ways of solving the problems of high-speed locomotion.

The skill of the racing driver is just one example of honing, to a high degree of perfection, operations embodied in very ancient neural circuits. They are always there, mostly working away under the threshold of awareness, which is why only cognitive science and brain research can tell you what is happening 'under the hood'.

We have already seen that there is rather more to perception than just registering the state of excitation of the different sensory tissues, and then experiencing the world in consciousness. There is rather more to motor coordination than just sending out a pre-packaged sequence of commands to the muscles.

The motor system does not only send motor commands to the muscles only, efference copy is used in perception; for example, to subtract the outcome of self-generated eye movement so that we experience visual stability in spite of constant movement of the eyes.

(Only those movements of the image on the eye that are *not* caused by a motor command and subtracted by unconscious inference is experienced as movement out in the world. Image motion that results from movements of the eye caused by motor commands is not perceived as visual motion. This is not all-or-nothing of course: if you do large, fast movements you do experience some residual visual movement. Try it.)

We are not usually aware of our eye movements or blinks that in most of our waking hours interrupt visual input several times per second. (Now that we mention it, right now you probably are!) This is because they are for our brain wholly *predictable*. Which means *they produce no prediction error.*

While the bottom-up sensory input signal from the retina changes massively, the relay stations bringing this information into the brain (the brain stem and primary visual cortex) do not report this massive change forward. This is an example of predictive coding in brain design. Your brain is 'wired for prediction' at all levels.

The predictable changes in the image (sensory information) are said to be explained away by the efference copy, and so the brain does not unconsciously infer motion in the world.

For all of this to work, however, the brain of course has to know what the predicted of sensory effects of all the various different types of motor actions are going to be. And for this, your brain has to *know how your body is shaped.* Which is the key to how the brain treats the car as an extension of the body.

BODY LANGUAGE

Whilst practicing a piece, ballet dancers will 'mark' the sequence of steps. In a systematic way, they perform the movements as mere gestures, not with the amplitude or the power of complete performance.

Not only does this reduce the exposure to practice injury, it may be positively helpful for learning. The relative ease of performing the marking, rather than the physically demanding movements of the dance choreography, may free up cognitive capacity – that is, attention and memory – for analysis, monitoring, detecting errors, and so on. This may help in learning the sequence.

For explicit communication and step-by-step conscious analysis, the steps can be marked with the *fingers* alone – projecting the movement sequence to a completely different set of muscles and joints!

The same thing can be observed in downhill skiers at the top of the hill, 'marking' a run with body movements. Or with synchronized swimmers projecting leg and body movements to hands and fingers when going through their program out of the water.

Aerobatic teams go through their programs on the ground as well, the pilots rehearsing the order and the vital synchronization of what everybody must do, and when, using their hand movements as a substitute for manoeuvers like banking, going vertical etc.

And racers explaining the reactions of their car or bike of course do the same, with a kind of symbolic language of different gestures for understeer, oversteer, weight transfer etc.

All these routines create a concrete analogue model of the actual movement, preserving timing and hierarchical and sequential relations. Relations between the target body parts (the marked movements) match the relations between source body parts (the gestures). The target choreography maps to the marking. The remarkable thing about the last two examples (pilots and racing drivers) is that it is not parts or movements of the performer's own body that are being mapped to hand movement – it's the dynamic behaviour of a vehicle!

There's a whole array of hand gestures and movements to describe what you are doing and how the vehicle responds that come with learning the 'sign language' of the racer. The race driving instruction in textbooks have quite little on this, though, as gesture does not lend itself to printed communication.

To a certain extent, a lot of driving textbook terminology is a kind of translation of engineering jargon to racer speak. There is also a relative lack of proper scientific understanding of the non-verbal aspects of communication (compared to the level of detail and depth that linguistics can pick apart language). It would be interesting to know more about difficult-to-verbalize aspects of perception and memory. How are they encoded into gesture and movement?

And even more interesting: how do these relate to the sense of space and the body schema of the expert– the way the internal models of the racer's brain keep track of eye-head-body-vehicle movement? Here's how I (OL) think it could work:

When You Become 'One' with the Car (As Far as Your Brain Is Concerned)

Proprioception is your sense of body shape, posture and orientation in space. It is based on signals from touch and pressure receptors in the skin, and tension receptors in the joints and muscles, balance organs in the inner ear, as well as motor efference copy. Kinesthesia, the sense of body *movement* in space, also depends on visual information (sometimes called exproprioception). These all belong to the job description of the same core systems, including parts of the attention networks in the parietal cortex, fed by the dorsal visual stream and receiving efference copy from the motor system.

They are doing coordinate transformations between signals arriving in eye, skin, muscle and world coordinates, maintaining your brain's internal model of where your body and its boundaries are, and how different motor commands will affect body movement.

This body schema is, however, highly plastic. That is: modifiable by experience. What I just said, in neurological jargon, is something really cool: your brain *does not know innately* what shape your body is, and how it can move. It has to unconsciously infer this information by integrating multiple (proprioceptive and exproprioceptive) sensory sources.[144] The brain will learn during development what shape of body it is controlling.

This makes evolutionary sense. With a plastic body schema, you can turn a squirrel into a bat by mutations that grow wings. You do not need to simultaneously get into place all the right mutations that grow a 'wing-command module' in the cortex, to go with the wings. If the cerebral motor system is multipurpose and plastic, then robust control is guaranteed.

It also makes developmental sense: the brain cannot count on a specific number of arms attached to the body, or a specific number of fingers on those arms, for example. Due to genetic reasons – or an accidental amputation – an arm or finger can go missing. (There is a genetic mutation that makes you grow six fingers, and the motor cortex appropriates this just fine. I am not aware of conditions that would make you grow four arms, Green Martian-style, though.)[145]

Perhaps the 'unintended side benefit' of this adaptivity, from the evolutionary point of view, is that the brain's design can take advantage of novel changes in 'body shape' that extend movement capacities. New mappings between motor commands, contact forces, and movement. Effectively it means that your brain can learn that *'Hey, I seem to have grown some wheels!'* All without need for complex, coordinated, genetic rewiring of the brain.

When you pick up a tool, such as a stick, the boundary of the body schema is extended outwards in space – you can now touch, manipulate and *feel* objects that are further away. You can feel the point of balance of the stick. You can feel texture at its tip. And note that we do feel the texture at the tip – we do not feel it at our fingertips. (For example, the texture of the stick at the fingertips can be hard, and quite different from the felt soft texture of the surface.) [146]

The stick is said to have been appropriated to your body schema. The stick has become – in a literal, neurological sense – a part of our body.

Another example of appropriation is how we cease to be aware of the feel of our clothes. Their predictable haptic (touch) responses and constraints on action are appropriated into the body schema. The clothes and the tools become 'transparent to awareness'.

I believe that this is also how a vehicle becomes an extension of the body: how a well-sorted car 'fits like a glove' so that you 'almost wear it'.

These metaphors in racer jargon probably point to the appropriation of the vehicle into a flexible body schema. At some point the vehicle becomes equally transparent to awareness as your clothes are. You and the vehicle 'become one'.

You are able to feel the tyres' loads and adhesion. Through the steering wheel, pedals and the seat, yes. But think of the way you feel the surface at the end of a stick: you feel the texture of the surface at the end of a stick. Through your fingers, yes, but at the tip of the stick.

Give your brain enough evidence that your body has wheels (seat time in a race car), and the brain constructs internal models to go with this new 'body shape'. It's quite amazing, really.

This of course requires that the vehicle itself behaves 'as one' in the first place. If its chassis and tyres and engine are not working in *mechanical* synergy, there is no way that the driver's *neurological* synergies can fully extend into it. Cornering in a car where the rear axle does not seem to know what the front axle is doing does not quite give you this kind of 'transparent' handling and 'oneness'. Right?

And if the vehicle does something unpredictable – the engine cuts out, you hit a patch of oil, something breaks and falls off or whatever – it ceases to be transparent. You zoom back into your own skin, as it were, and become aware of the vehicle as a separate entity.

Again, we have the predictive brain design working here. The mismatch between the predicted movements and feedback (proprioceptive and exproprioceptive) is being passed forward, until it reaches the levels of processing associated with awareness.

LIMITS OF SPOKEN LANGUAGE

This goes also some way towards explaining why it is so difficult to put what you are doing into words. Why it is so impossible to communicate the intricacies of 'feel' and control to non-experts.

The ballet dancers are marking the steps with their fingers, not with words, for a good reason! The neurological reason is that the language system is quite a different system from the body schema system (**Figure 8-1**, overleaf).

The language system, which resides usually in the left hemisphere, operates by taking lexical elements (basically, words) and arranging them into syntactic structures (basically, sentences). The body schema system on the other hand, operates by taking multiple sensory, motor and spatial memory signals, and arranging them into action and awareness. They trade in quite different sorts of information.

An analogy is helpful here to illustrate the distance between language and skill. In terms of the way driving skill is organized, it is similar to playing an instrument.[147]

Here is one way one *could* describe playing two chords on a guitar:

'With my index finger on the third string of the first fret, and my middle finger on the fifth string and my ring finger on the fourth string, I am strumming all six strings. When I now come up to the fifth beat, I move my hold from the first to the second fret, and I'm keeping the index finger on the third string – but at the same time I'm switching the index finger to the fourth string and my ring finger to the second string. Maintaining the beat, I nevertheless switch strumming to only strings one to five, while making absolutely sure that I always miss the sixth string.'[148]

The above description certainly is not a good description of what the skilled musician is consciously thinking about when playing!

Nor is it by any means a complete description of *all* the necessary (procedural) knowledge involved: the details of rhythm, intensity, and timing that the player needs to get right to make the music flow. Yet this sort of knowledge must be there, implicit in the player.

The explicit description captures some aspects of the skill, by re-encoding them into words. But actions use a different code: motor programs used in procedural knowledge, not lexical concepts used in explicit knowledge. The verbal description, then, is just a means to communicate some small part of the information. For example, when you need to explain the basics to a beginner learner. Lexical information is, literally, 'just words'.

To develop the skill, and to develop the knowledge *how* to play the guitar, even the beginner learner cannot rely on this explicit verbal knowledge alone. (Just try giving someone the above description on a

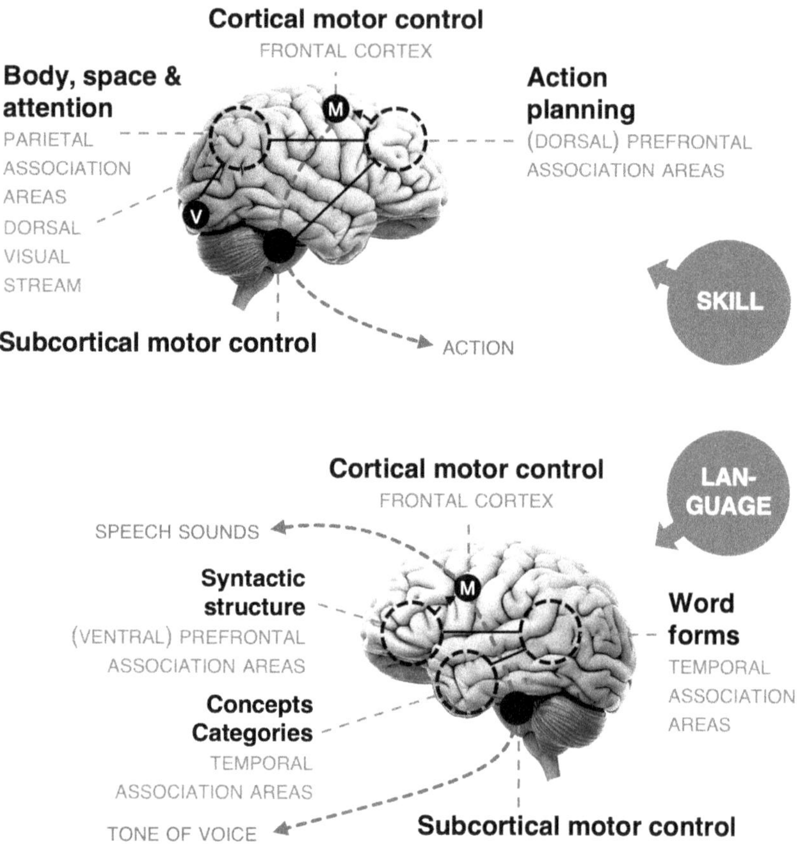

Figure **8-1.** Skill and language reside in largely distinct brain networks, even though the final motor pathways are the same (all systems must control the same muscles, after all).

The frontoparietal networks for attention, body schema, mental imagery, and sense of space (top) integrate perceptual information from the dorsal streams. This allows the expression of procedural knowledge in skilled body movements.

The temporal and frontal language networks (bottom) integrate conceptual knowledge, and ventral stream perceptual information into word and sentence structure. This allows the expression of declarative knowledge in language.

This is why it is not always easy to put what you are doing into words… or to put your thoughts into action…

piece of paper, and a guitar, and telling them to have a go at it!)

For one thing, you must also *show* them what you are doing, not just say it. It would be impossible to learn to play the chords from the above description if you were not at the same time shown what to do.

One way racing coaches do this is to sit in the passenger seat and give the learner the guidance information their own brain cannot yet supply. *'Brake brake brake, steer now, throttle on, more power, more power, FULL power!'*

Also, the learners absolutely must also try it out themselves. And try it out many, many, many numbers of times! This is such a pervasive feature of humans learning any skill whatsoever that it may seem too obvious to point out. But the reason why it is so universal – true of every real-world human skill on the planet – lies in our brain architecture.

The way memory in the skill-forming part of the brain is organized – which is different from the way memory in the verbal, thinking, part is organized – means that this system just cannot pick up things in one shot.[149]

It can't 'get' what needs to be done (at the control level) without a lot of repetition; a lot of actively performing actions and observing their outcomes, so that the control can be corrected based on feedback. That's just the way the brain is wired up.

Which is probably a good thing – learning by chipping away at it, one action at a time, small variations on the theme... this may be a failsafe to prevent us from trying out something completely different... and catastrophically wrong. Language is different: you can repeat a sentence on one hearing. (And you can take it from me: it is very, very easy to say something catastrophically stupid.)

Computers are organized more like language; so computer programs do not have this failsafe. You can upload a tune into your iPod, and it will be able to play it instantly. But get just one bit wrong and it can all go spectacularly sideways. (Like switching 'full brakes' into 'no brakes'... just one bit...)

This difference, between knowing-what-to-say systems vs. knowing-what-to-do systems also underlies the difficulty of expressing skill in words. To the uninitiated it is just so much gobbledygook; and even for experts who, unlike naïve observers, can 'get' the meaning from verbal cues or miming, there is a limit to communicating the nuance.

Think of how a guitar player might improvise; using certain scales, playing around with timings and particular note choices... it's all happening at such a speed, and in systems that run parallel and much faster than the verbal/thinking parts of the brain, that a player can't consciously describe the process.

Thousands of intricate decisions are being made – the little stretched and bent notes – at a level of understanding that is a bit mysterious and hard to nail down. That's what race driving is like, too.

Although language or speech are often heralded as the ultimate system for the cultural transmission of knowledge and technology (especially by scientists who study language!) there are severe limits to what can actually be effectively encoded and communicated by language. (At least in the absence of showing and mimicry, and a shared physical environment to visually point to.)

Along with language, mimicking others is something humans are uniquely good at, and have an appetite for – other apes are much less prone to 'aping' each other than humans.

You learn by unconscious 'osmosis' by watching others perform – not just by doing what they tell you to do, or what they say they are doing. (Remember Stirling Moss's brake marker in the tree at Arnage...?) This is also the basis of some forms of coaching where you are asked to visualize taking a corner as if some role-model were driving.

ARE HUMAN BEINGS REALLY DESIGNED TO GO THIS FAST?

It is not speed itself that creates a challenge in racing, but the consequences of it.

One consequence is kinetic energy, which goes up in proportion to the square of speed. This is what affects your braking distances and the severity of your accident – and so the danger involved.

In shedding that speed, physics say the kinetic energy absolutely has to go somewhere – like heat dissipated to the airstream from the brakes and the breaking of molecular bonds between pieces of rubber when you are leaving skidmarks... or to pieces of your car flying around in a crash, which retain some of the kinetic energy. The rest goes into breaking molecular bonds in whatever you hit, and in your car's crash structures – and, for obvious reasons, hopefully as little as possible in your body and brain. These are consequences engineers responsible for car and track design have to deal with.

Other than being suspended in a couple of millilitres of cerebrospinal fluid, which gives nominal damping on impact, our brain is not very well designed to cope with shedding speed in a collision. It's encased in very, very thin skull bones (the bone on the temples is so thin it's almost translucent), and the skull bones at the front actually have sharp edges on the inside, and so they easily do damage to the brain on frontal impact. (Which is why boxers who must endure frequent frontal impacts are at an increased risk of developing chronic frontal lobe damage.)

What is more, unlike skin, bone, cartilage or vessels, damaged brain tissue does not grow back. This is a big deal. If you sustain damage in the brain, it basically stays there for the rest of your life.

So, really, the brain is not designed to withstand hitting anything at much more than walking pace. So, in that sense you may say humans are not, biomechanically speaking, designed for the kinds of speeds we seem to enjoy so much.

However, another consequence of speed – more pertinent to analysis of cognitive operations and skilled performance – is the rate of information processing required. The *rate of events* that need to be registered, discriminated, interpreted, and responded to goes up as speed increases. Conversely, the time available to make a decision gets shorter.

How well designed is the brain for these more cognitive consequences? Now we are talking about the design of the brain in terms of its information-processing architecture. And as it turns out, the question of how well the brain copes with speed now shows up in a very different light.

The necessary rate of information processing is actually not a function of *speed* as such. For example, I (OL) am writing this now in an aircraft moving at some 800 kph – but the cognitive load, the rate of processing required from the pilot, is probably minimal. For all I know she, too, may be having her in-flight meal.

Much more demanding is landing, where the actual speed is much lower, but the *proximity of the ground surface* means that the rate of events is higher. And put a driver-pilot in an 800 kph land-speed record car, just millimeters from the surface, and the task demands go off the scale!

The general point here is that information-processing demand depends on the speed together with the spatial scale of the environment. It is these together – it's their ratio! – that determines the necessary rate of information processing[150]. Remember, this is a big part of what makes Tabac a quick corner. It's the proximity of the barriers relative to the speed at which it can be taken.

Now, an opinion I sometimes hear voiced, especially by experts in traffic safety, is how our brain 'did not evolve for' the kinds of speeds modern vehicles take us (and why, therefore, we all need to be regulated and controlled by experts in traffic safety – or hand over the driving task to machines).

One may legally reach the heady speed of about 70 mph on the roads of most countries (in Germany, even more, long may that continue). As far as complex information processing goes, this is not really a huge strain, however. Indeed, people have been known to talk to cell phones, eat lunch and do other things while at it…[151]

What the focus on speed alone fails to take into account is that the highway environment is *designed* in such a way that as speed increases, the distances to anything you would need to react to, or might hit, increases as well. Therefore, the time to react, or rate of new information for the brain to digest, does not become very high. Indeed it may go down. (At least as long as there are no stray ungulates wandering into your path, for example.)

It's not like moving from walking pace to running in a dense forest, for sure! In terms of information processing rate, it's probably not even on a par with walking pace in the forest… And for those things we are surely 'evolved for' in anyone's books, wouldn't you say?

So, just what did evolution equip us with to be able to do it?

THE PREDICTIVE BRAIN IS DESIGNED FOR SPEED

We so easily think that we perceive what our senses tell us. This seems almost a point of logic. Is it not almost the very definition of perception that it's the acquisition and interpretation of sensory impressions? How could it be otherwise? This view can be reinforced by looking at the physiology of sensory encoding and the discussion of reflex actions – like we had back at the beginning of **Chapter 6** – which is what most textbooks start with.

Your eye registers light, it sends a signal to your brain. Then your brain (in some magical way) transforms the information it is being fed into visual experience. Or: the doctor's hammer stretches a tendon in your knee, and the sensory input triggers a reflex. (This time without conscious experience; the sensation of the tap and of the movement of the leg emerge in the brain later, as the relevant signals travel up the spinal cord and find their way into the brain's proprioceptive system.)

But this picture is misleading in making the jump from the signal arriving in the brain to perception, at least insofar as it tacitly, mistakenly, suggests that the information that generates percepts would be provided by the sensory signal... when in fact, most of the information needed for perception already resides in the brain. As does most of the information needed to coordinate motor actions. But not all of it.

The brain is wired for prediction; this is what gives it speed and efficiency. The sensory system uses predictive coding: feedback is there just to check for, and report back to the brain, any *prediction error* that allows new information to be encoded. But the perception and action flow from the internal models (long-term procedural memory), which is what allows it to be fast and efficient.

And perception is not merely a response to sensory input, it is a process where the brain predicts sensory input, before it arrives, and then checks the predictions against the data. (And learns and adapts its internal models, so that next time you will expect, and perceive, something different.) The sensory information is used to fine-tune the predictive machinery. We perceive not just what our senses indicate is happening, but what our brain *believes* is happening, based on prior observations and memory processes.

This is particularly useful when things are happening *fast,* as there is a physical delay in the transmission of sensory information that cannot be overcome. These delays are caused by the physical properties of sensory responses, neural transmission and muscle contraction.

Consider a baseball pitcher at the split-second he is about to release the ball. The point of release must be timed to within a millisecond. But it takes about five milliseconds for muscle commands to travel down the arm. It takes about one *hundred* milliseconds for optic information encoded at the retina to reach the brain.

So, it is clear that timing of the release cannot be performed in response to current optic information – the most current available optic information at the time of release is already well over 100 ms old (transmission time plus whatever time it takes the brain to analyze the information and adjust motor programs). Indeed, given motor delays, in the milliseconds before release the muscle commands specifying the release must already be well on their way down the forearm.

These kinds of physiological observations led psychologists and physiologists in the 1970s to come up with the hypothesis that instead of reflex responses (which are fundamentally limited by reaction time delays) very fast and accurate, well-learned movement patterns are driven by ballistic motor programs or 'precognitive' responses.[152] Pre-packaged motor routines that once launched run their course automatically and without further effect from sensory/perceptual or memory processes.

Sensory information and memory would only be used to select from among possible motor programs and set their parameters, including the exact moment of launch, after which motor systems take over and the motor program mechanically runs its course. This type of anticipatory control is pure model-free feedforward control, as there is no role for feedback information from the senses after the action is launched.

There is a snag, however: when the complexity of the motor action increases, the time it takes to complete the action grows longer. If you need to pre-package the *whole* action sequence into this kind of motor program, then you need to launch the program earlier and earlier and earlier in order to make it run its course, *and* reach the critical point (e.g. ball release) at exactly the right moment.

Indeed, the timing of motor program initiation itself would need to be as accurate as the desired outcome needs to be. And the variability in the program or its movement parameters (especially duration) should be very, very small. Otherwise it would not make a difference if you initiated the motor program at exactly the right time, as variability in the duration of the movement would cause the end-point to spread. But achieving such accuracy in the initiation time and exact duration of the motor program is not physiologically realistic.[153]

But really, these sorts of motor programs would only be needed if the brain were trying to replicate complex actions in machine-like, metronomically precise repetition. But you know already this isn't the case. You are not doing things wrong if you don't always hit exactly the same apex, as long as your lap time is there!

As we saw in **Chapter 5**, measurement of actual complex movement sequences by skilled humans (blacksmiths, triple jumpers) has shown that in fact there can be considerable variability in the timing and in the exact pattern of a movement, without affecting control. Careful video-based analysis showed that there is more variability in the arm swing directions and joint angles, than there is in the direction and speed of the hammer's contact to the metal. The outcome variability is smaller than the variability in the actions leading up to it.

This tells you there must be more going on than ballistic, fixed action patterns – that feedback must be somehow incorporated into the performance. Especially so for complex action sequences – like shifting gears or turning into a bend at speed. Because of their complexity, super-accurately pre-programming them is difficult: there are so many places for random errors to creep in and accumulate.

But on the flipside: the longer the duration of the sequence, the more there is now time for some relevant feedback information to make it to the brain during the action, and be incorporated into the unfolding action (or action-prediction what we called 'internal simulation' earlier).

The question then becomes: how does the brain use sensory information as feedback to guide fast-paced actions that are too complex to be ballistic motor programs, and happen too quickly for reflex control?

For most complex tasks, including driving, the details are not known but the general picture has begun to clear up since the end of the 90s when computational neuroscientists began to challenge the view that skilled action is generated as *either* 'preprogrammed' feedforward motor commands *or* as purely reactive feedback loops.

Predictive processing gives a modern approach to the problem of feedback delays,[154] and the predictive processing framework[155] can give a unified treatment of perception, action and cognition.

To recap the core idea: functioning in dynamic settings is based on continuously predicting the perceptions arising from states of the environment. Action depends on internal models: generative models predicting the consequences of actions. (The consequences could be perceptual inputs, or future events, or other things not currently observable of the environment).

The accuracy of the predictions is consistently tested via physical interaction: actions that generate feedback. Prediction errors (sent up via predictive coding) are used to adapt the internal representations, and for motor control.

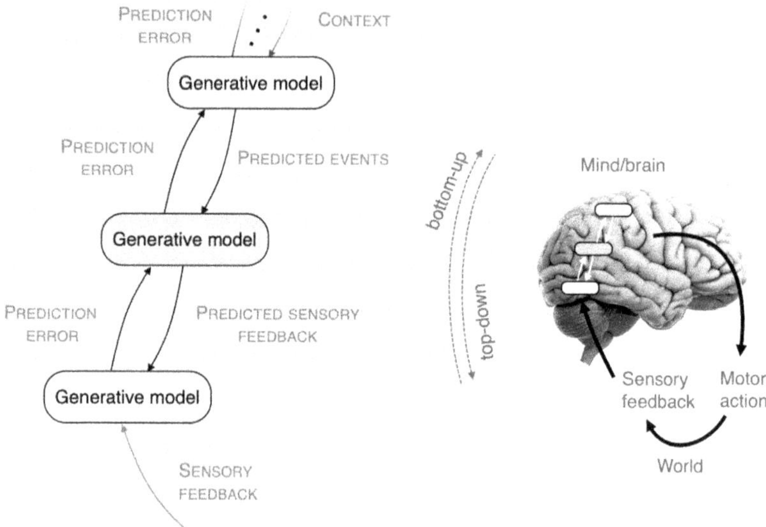

Figure 8-2. The predictive processing view of the brain, as a hierarchy of generative models. (Generative simply means that the model can generate predictions). Specifically, predictions of their own inputs, which it can then issue downwards in the hierarchy, so that the lower level can evaluate the prediction and report back prediction error.

The important thing here is the difference from *both* (i) pure feedback control (where action could only be initiated as a response to a change in stimulus, viz. the difference between a reference value stored in memory and the measured value of the controlled variable) *and* (ii) ballistic feedforward control (where the motor program cannot be updated by new information that becomes available from sensory signals or from memory after it has been launched).

In predictive control, action is driven not by error calculated from feedback after it arrives (with a delay), nor by a prepackaged program,but by these two things:

1. A generative internal model in memory that represents the current or even future values of controlled variables, or values of hidden variables that cannot be observed. (Then it's called model-based control.)

2. Feedback prediction error itself. (Then it's called, somewhat obscurely, active inference.)

The difference from ballistic feedforward control, where the releasing stimulus specifies a complete movement sequence, is that in predictive control the brain still uses feedback: to generate and maintain an up-to-date internal model of the situation. This is then used for motor control.

The difference from feedback control is that you are using the brain-internal information, not the delayed feedback information, to control the response.

Because the brain is acting on the basis of predictions (from internal models), it can control actions with zero reaction time, even negative reaction time. Think about it: if the brain knows *when* an event is going to take place, it can start preparing for it already well before, and it can time the response to the predicted occurrence.

Compiling these models is slow: it happens over the hundreds or thousands of hours of seat time. But the more accurate the models, the finer the driver's sense of 'orientation' and her 'feel' – and the greater her powers of anticipation. And because memory retrieval is fast, the models can be readily called upon at speed.

By being able to anticipate, your brain is 'giving itself more time' to decide how it is going to handle said event: the choice of action is no longer limited between the time when the event becomes observable (because of delays this is always going to be some time after the event), and when it is too late to avoid its possibly untoward consequences. At racing speeds that time window would be all too brief to come up with complex or finely tuned (re)actions.

A good example is catching a skid. Drivers and TV commentators often talk of 'good reflexes', but true reflexes are reactive behaviours, not anticipatory. Closer to the truth is the metaphor of being 'on top of the car' or 'ahead of the car' – managing or even preventing skids by minor anticipatory adjustments (vs. 'falling behind', when the responses are too late or too slow, and the skid has time to develop into an out-of-control tank slapper).

A typical beginner error is to start steering into a skid with a big, slow countersteer, and then, when the front wheels get a grip, *violently* spin the other way. According to driving technique manuals you should steer into a skid until you observe (no – predict!) the yaw rotation coming to a stop, and then quickly reverse steering lock to prevent, or be ready to catch, the evil counter-skid.

If your brain does not have a good idea of when the skid is going to end, it is not going to be able to cope with the speed of the counter-skid – that is, if your brain *actually* waited until it could visually observe the first yaw motion stopping, your counter response would probably be too late… and so you'd get caught out.

It is building up these sorts of predictive internal models about what the car is doing (and what it is going to do) that underlie car control skill. They allow the experienced driver to 'stay on top of the car', and not fall behind.

Sometimes, when the process really clicks, the driver feels like time itself is slowing down. There is all the time and space in the world to move the controls, the scene is not rushing at you in spite of the speed. You get 'in the zone' and you achieve flow.

The power of this information-processing view of the brain is that the same basic architecture is repeated at all levels: computations across different levels of temporal scale, abstraction, and complexity are fundamentally all doing the same kind of predictive processing. This gives a unity to the view of the brain.

And if this theory is on the right track, then it has some interesting consequences for understanding how the brain uses feedback, and even actively seeks to create it. Feedback is the currency that all learning and skill development trades on. That's why it is so vital. And prediction is vital for feedback. Consider this: the predictive processing view says that for most of the brain, 'feedback' is actually prediction error. This emphasizes the active role of the brain. After all no prediction, no prediction error. And if feedback *is* prediction error, then simple logic yields the following argument:

FEEDBACK = PREDICTION ERROR

NO PREDICTION → NO PREDICTION ERROR
Therefore (by substitution):

NO PREDICTION → NO FEEDBACK

What this means is that the hazier the predictions the brain makes (the less detailed the internal models), the fuzzier and therefore less potent – less meaningful – the feedback will be. If the prediction is not very strong to begin with, then comparison to actual feedback can only produce a fairly weak feedback prediction error, leading to only sloppy corrections, fuzzy recollections, and limited update of internal models in memory. Not much learning will occur. Example follows:

When I (OL) went to try the GTS-RS simulator Alan worked at, I of course very quickly had a spin. Two almost identical spins, in fact. (Embarrassing, but nicely illustrating the point above about limited learning.) The interesting bit was that both times I had to ask Alan what happened, and both times he said I'd been too slow to correct the steering after the initial countersteer to a skid. Because my brain was lagging so far behind what the car was doing at that speed (which if I'm honest was not even very fast...) it was not able to log any detail of the event. I just suddenly found myself in the wall but how I'd ended up there was a blur.

This point of view may make sense in your personal experience as well. If you've had a *predictable* spin or crash – this happens when you are taking a deliberate risk and your brain knows full well the potential consequences if you cannot pull off the manoeuver – then everything is logged to memory down to minute detail. You remember the crash as if in hi-definition slow motion. This is because your brain has a detailed picture to compare to the incoming feedback, and it updates this knowledge with the prediction error signals (obviously you predicted, confidently even, that it would be okay). But when there is an unpredictable incident – a sudden tyre blowout or someone rear-ends you... – you may have no idea what happened. You must go analyze the data and reconstruct the course of events from video, *'because it all happened so fast'*. Right?

Also, it is well known that drivers tend to prefer cars that act predictably. Cars that are 'faithful' and give clear messages about what they are about to do. But what makes a car 'predictable'? Prediction is not only a property of the car – a thoroughbred race car is *only* predictable to the expert driver. An everyday driver is sure to get a surprise or two! Predictability is dependent on the particular predictive models an expert driver has developed, over the years, when they've learned to drive. For a beginner who is lacking the internal forward models even the most predictable racing car will not be predictable – and will snap into oversteer and spin.

And while all drivers prefer cars that act predictably, not all drivers will prefer *the same car,* or the same set-up. It's not that uncommon for drivers to try set-ups and cars of other drivers and be horrified at the handling. This variation arises because not every driver has the same technique. So, when the preferred motor programs executed are different, so is the feedback the driver will get, and consequently the resulting forward models in memory will be different.

In short, what is predictable and what clashes with prediction depends on the driver. As a result, the perception and the phenomenology of the driving experience is different as well.

It is well known that the only way to achieve control at the limit is seat time. And what seat time gives you, under this theory, is the feedback prediction error your brain needs to update its internal generative models… at all levels of the hierarchy.

Because from this perspective, all the things that we have talked about in this book, all the building blocks of the driver's skill – the things that go in the 'navigation box' in the 'guidance box' and in the 'control box' – can be considered generative internal models in one integrated hierarchy. (Because, of course, your brain does not *really* have boxes inside.)

Knowing the track is the highest level: cognitive maps are very stable, enduring models that incorporate knowledge of the unchanging aspects of the physical layout. (What the track is like in different locations.) The chunks of reference points are really generative models, too. Also at quite a high level of the hierarchy. They are issuing predictions of what the bend will look like behind the blind entry, for example; what kind of guidance level information is expected to appear in the visual field.

During visualization, what happens is that you run these models offline: you activate the higher-level models and they initiate a downwards cascade to generate visual experiences, top-down. Keeping your eyes closed reduces sensory inputs that could conflict with the model (thus keeping prediction error down). And if you make the appropriate movements and sounds, physically enacting the actions, 'marking' the track like a ballet dancer, you are actively providing the system with some prediction-consistent efference copy.

In actually controlling a racing car it's all about generative models as well. The body schema and how it extends to the car (appropriates the car dynamics) would also be an example of generative models. The racing driver's brain is predicting *where* the car will go (in space, and dynamically speaking), and predicting its own motor commands. For example *when* and *what* brake, steer, shift or acceleration commands are going to be leaving for the muscles… and to the perceptual system as efference copy.

These generative internal models house your curve prototypes, your reference points, your waypoint chunks, your procedural knowledge of timing and control. They span complex, distributed circuits that spread through the cortical and subcortical motor systems, the parietal attention networks, the visual scene analysis and wayfinding circuits, and more. Within the living physical structure of the brain, they embody the skill of the racing driver.

Chapter Nine

Beyond this Present Horizon

WE ARE NOW COMING TO THE END of this particular journey into the racer's brain. We have looked at what happens in the mind and the brain when you go fast. We have broken down the skills involved into three categories – navigation, guidance and control – and analyzed the cognitive processes involved in each and the predictive brain mechanisms underlying all of them.

In so doing, we have introduced a new way to see the performance of the racing driver; the point of view from cognitive science. We have tried to do it in a way that would be interesting, fresh and unique.

Still, many interesting things have been left unsaid. When you look into it, there is a depth to this field – as there is to any mature field of human achievement – that a single book cannot hope to cover. As with most things that appear simple, once you scratch the surface there is complexity and detail.

Like, kicking a ball is simple, right? Or swinging a bat or a racket of some sort; that's simple, right? Then again, returning a 100 mph serve with millisecond precision, reading the game, feeling what your body is doing, anticipating your opponents and understanding the ebb and flow of positioning, all the feigns and the countermoves, and then picking the right line of play… this is far from simple.

Motorsport is the same. There's more going on than just turning the steering wheel and pressing some pedals! And on many topics, we have indeed only scratched the surface. We have gone through a lot of theory, but while '*In theory, there is no difference between theory and practice. In practice, there is*'.[156] We have not really looked at all into specific driving techniques; what works and what doesn't work in the real world. Nor have we looked at the biomechanics of the driver-vehicle system, and how they are constrained by, and exploit, the underlying physics. And while we have discussed some cognitive aspects of skill development, the question of talent, and the role of 'genetic endowment' in skill acquisition, if any, was left untouched.

There is little discussion of the problem solving that goes into troubleshooting car setup, or the way the best drivers develop and troubleshoot their own driving. And, as you may have noticed, we've really only concerned ourselves with the skill involved in driving a racetrack – we have not even begun to analyze *racing* against actual opponents!

And we have not really mentioned, not even in passing, things like emotional maturity, leadership and team building, or the way the best drivers are constantly identifying their own strengths and weaknesses: figuring out ways to address their weaknesses, and capitalize on their strengths. All of which is so, so important for developing a successful career, in motor racing as in many other walks of life.

And then, every driver is different. We have on purpose emphasized the universal aspects of human cognition and the general design of the human brain. But each individual human brain is unique in its micro-wiring, shaped by experience. No two are exactly alike.

So, there is probably no single optimal way to achieve performance (in any complex domain of skill and cognitive ability). What you are optimizing is the interaction of your unique brain and physique with a complex environment; and when you go into the details, what is the best way to get the job done *for you* may not be the best way for someone else. The stopwatch doesn't lie. Well, yes, but different drivers can achieve near identical lap times with very different styles, and where on the track they make the time can vary.

So, what gives an individual their particular cognitive and emotional 'style' – and how competitors play on their own and their opponents' strengths and weaknesses, here there's always likely to remain a bit of mystery.

After all, if the basis of some ultimate human achievement was simple enough for others to understand, maybe it would not remain the ultimate achievement for very long...

Figure 9-1. Motorsport can be an intensely personal and solitary affair. The driver's actions may be recorded in vehicle telemetry and analyzed by engineers in minute detail… but in the moment of action, what goes on in the solitude of the racetrack is an individual, personal experience. It is just the driver out there, putting it on the line, alone and unaided. Something experienced only by the few who go there.

Where do we go from here? Science now understands more about the mind, brain and behaviour than it has at any point in human history. Brain imaging devices are chugging away in countless labs, to render visible the metabolic activity of a living brain. Portable measurement devices, like eye-trackers and mobile EEG, allow scientists to study behaviour and the underlying physiology in the real world, outside the confines of the laboratory. Computational modelling methods allow us to model and simulate complex systems, at a level that was unimaginable just a few decades ago – even a few years ago.

But we are still far from a thorough understanding of even some of the most fundamental aspects of real-world performance. There are deep uncharted waters, with most of the interesting discoveries still waiting to be made. And much of that progress will have to come, not from new technologies, not from new measurement equipment, not from more powerful computers, but from new ideas. New ways of thinking. New ways of posing intelligent questions to nature. Discoveries in science tend to come from *figuring out what are the right questions to ask.*

More powerful measurement techniques, and more sophisticated analysis methods are then useful, but in themselves they do not solve the problems. Their complexity can even lead to a fragmentation of research into ever more specialized subdisciplines; with research in one field becoming more and more impenetrable to those working even in nearby areas.

Being able to bring together the insights and knowledge of scientists in different specialties (who are nevertheless working on aspects of the same problem) requires a multidisciplinary approach. In the particular case of understanding the racer's brain – or understanding the brain of the elite athlete more generally – these specialties would include perceptual psychology, cognitive psychology of expertise, brain imaging, cognitive and systems neuroscience and neurophysiology, and also artificial intelligence, complexity science, and numerical simulation in physics and engineering. (**Figure 9-2**, overleaf). That's a lot.

And these disciplines need to be brought to bear in a unified way. I (OL) for one believe cognitive science, especially computational cognitive modelling, has potential to do this.

In the case of the motorsport athlete, this approach complements racecar systems engineering and human performance engineering, and I hope that this little book might inspire some people in racecar and systems engineering to take a more cognitive approach to their subject matter.

Equally, I would like to see (dare I hope?) that psychologists and neuroscientists would begin to look favorably on driving, and other forms of real-world performance, as fields of study. To look upon the problem of genuinely understanding some complex human behaviour in a concrete real-world task setting not as merely 'applying' hypotheses developed in laboratory contexts, but as a necessary means of revealing something of general interest. *Only* looking at real-world behaviours really forces you to think about how different cognitive processes and brain mechanisms *integrate,* in a concrete way.

The biggest hope for progress lies in the creativity of some of you readers – especially students in any of the relevant fields – coming away with ideas, concepts, methods, or techniques that people have not even begun to think about yet.

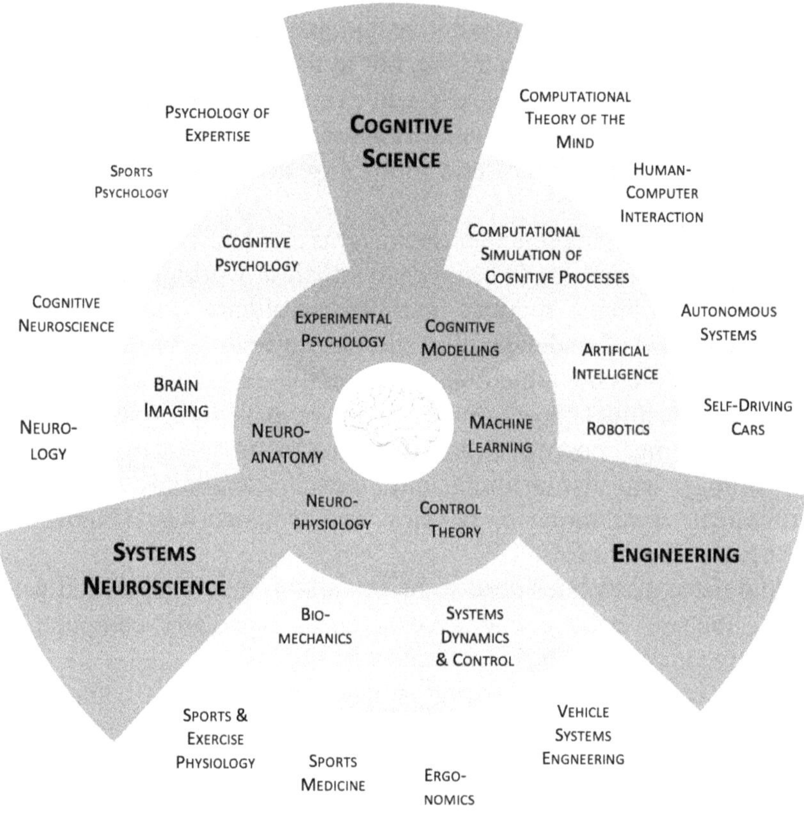

Figure 9-2 A full picture of human performance in race driving requires the confluence of multiple traditional disciplines, working on different aspects of it. This can only be achieved with unifying concepts and methods. And you do need to look at 'genuine slabs of human behaviour'[157] as well, not just fragments of behaviour that it is convenient to reproduce in the lab.[158]

If you are a student embarking on a career of experimental research on human behaviour and brain function, then, whatever your personal hobbies and interests might be – be it music, cooking, painting, bicycling, rock climbing… *indeed, driving!* – my advice would be to anchor your scientific thinking in your domain of interest.

And not just anecdotal connections to the domain, either. Not just the established perspective of fixed concepts and theories, within your scientific specialty… or even within your field of science. Try to understand *everything about it*, from as many scientific points of view as you can. The physics of it, the chemistry, the psychology, and cognitive science of it – even the economics and cultural and natural history of it. (You can always do it to some approximation.) Your general scientific thinking will improve as a result.

If you are an expert in race driving, driver training, racecar engineering, sports psychology, cognitive neuroscience – or in any of the numerous fields of science this book touches on – you may not agree on everything we have had to say. That's okay. While we've tried to be as factually accurate as we can, too many of the facts are just plain unknown. (And our two brains, combined, only carry around so many facts – or even opinions! – to draw on.)

If you disagree with anything, then, please do go and do the research, and *show* us where we were wrong! If just one person who reads this book goes and does an experiment that falsifies a wrong idea, then this book will have been more than worthwhile.

And if you are not a racer, maybe not even a race fan, maybe this scientific look at the racer's brain has yet been of interest to you, has entertained you, has informed you. If we have been able to share just some of the fascination we have for this subject, if this work has given you some insight, or piqued your interest about the brain, or about racecars, and about how these come together, then we have achieved the purpose we set out for ourselves.

Our modest ambition in this book has been to bring into the literature insights, observations and ways of thinking heretofore not present in the popular or in the technical works on racing. We believe that based on our backgrounds and expertise we have an interesting viewpoint to offer on 'what motor racing is all about'. We wanted to share this viewpoint; and if you have read this far, you must have found something of value in it as well.

We hope you have found not only new answers to questions you've always wanted to ask, but, more than that, discovered questions you never thought to ask – this is the beginning of knowledge.

Epilogue

'IT IS NOT THE TITLE THAT MAKES A CHAMPION, BUT WHAT HE HAS GOT IN HIM.'

– PIERO TARUFFI[159]

IN 2009 I (AD) was a plucky young motorsport reporter for my own karting website. Karting has always been my passion, as I just don't think anything else comes close to the all-round experience. From the accessibility to the simplicity to the range of driving experiences which offer varying degrees of purity and performance – from an 8hp Indoor kart accessible to everyone all the way to a supercar-destroying SuperKart monster. Ayrton Senna described it as the most breathtaking sport in the world, and in my view he wasn't wrong.

This passion is shared by many other notables in motorsport, including the great man Michael Schumacher… and what many fans of motorsport might not realize is that you can often find some of the world's greatest drivers smashing in a few laps at their local kart circuit. Schumacher, even in his prime years at Ferrari, was often at the kart track.

Most famously at his local kart track in Kerpen, where he competed in a round of the 2001 CIK FIA World Karting Championship. Taking advantage of wet conditions (in the dry he wasn't massively fast) he scored a brilliant podium behind Sauro Cesetti. He beat Lewis Hamilton and Nico Rosberg in that race, who'd both go on to be World Champions in F1. Vitantonio Liuzzi was the overall champion that year. This performance was indicative of Schumacher's general approach to racing. He appeared to be at complete ease to put his reputation on the line to compete with comparative 'nobodies'. (They're certainly not nobodies to me, though. Top karters are among the world's best drivers.)

Wind forward to 2009, and rumours abound that Schumacher, by then the most dominant driver in F1 history, is making a comeback to Formula 1 with the newly formed Mercedes-Benz team. (The team then went on to become a juggernaut in the hybrid turbo era, and one of the most dominant F1 teams in history.) At this momentous time, where could Michael be found? Well, he was at the Rio Hotel's car park in Las Vegas, readying himself to compete in the SKUSA Supernats.

The Supernats is North America's premier kart event. The States are somewhat different to Europe when it comes to all things motorsport. Tom Kutscher, the owner of SKUSA, does an awesome job of attracting some of the world's best drivers to his event each year, and the Rio Hotel car park is certainly a very different affair than what the euro circuit regulars are used to.

The whole event was like one big head trip to me. I wasn't used to this kind of experience. Karting to me usually meant long weekends spent in a field in the middle of nowhere, and you consider yourself fortunate if you have clean toilet facilities. This was a whole different ball game: roulette tables, endless pinging slot machines and the city that never sleeps. And I was sharing it with Michael-bloody-Schumacher!

So just how did Schumacher get on? Well, in the end he slotted home 7th, though he was helped by a few drivers coming together ahead of him. Bas Lammers was in fine form that year and took the win. His driving was just simply a step up from everyone else.

The fortunate part for me was that I had privileged access to the circuit. In karting, photographers and media can get right up close to the action. I was able to observe up close pretty much every lap Schumacher put in during that whole event. He was flying past me by inches, at speeds over 70mph… The kart he was driving was a 125cc 6-speed gearbox kart with front as well as rear brakes. These karts are absolutely brutal machines; without proper rib protection you might not be getting out bed the day after driving them, because you'll have broken ribs.

What's great about karts is that the driver is so exposed, from a driving perspective. You can very quickly assess the priorities a driver has, and interpret what you see. A few things stood out. For one, Michael wasn't the most subtle driver. He liked to drive 'on the nose'. He was heavily reliant on the front and would 'vee' the corners off quite aggressively. It worked fantastically when the karts first hit the track on Thursday morning, when it was green (and car parks are as green as they come). He went fastest. As the track evolved, he started to slide back a bit.

But it was not the style itself that really made an impact with me. Nor the results. (Those who know karting know that a Top 10 result is fantastic, and it's not a slight on Schumacher that he didn't win – karting is insanely competitive.) It was Schumacher's *approach* to racing. He was moving into his 40s at the time. He was a seven times World Drivers' Champion in F1. His net worth was in eight figures. But here he was testing himself against some of the world's best karters, and he was serious about it. This much was evident. I recall one instance when he went side-by-side with an American driver through an extremely fast left hander where there was zero run off (it's a car park with barriers, remember). Tyre smoke bellowed as he turned into this guy. His commitment was *through-the-roof!*

The take-away from this: even when we lay aside the technicalities of line, speed and load transfer management, and all the other stuff... there remains an element to a driver which makes a difference. Schumacher certainly had unique aspects to his driving technique, which may have contributed to his success – but it was in that car park back in 2009 that I learnt up close *what champions are made of.*

ACKNOWLEDGEMENTS

Many people gave of their time to discuss with us and share their insights on the topic of racing and/or the brain, contributing directly or indirectly to the maturation of this project. Some colleagues commented on drafts at different phases of this project. All discussions and feedback improved the final product. So, thank you (in alphabetical order):

Jani Backman

Ross Bentley

Karun Chandhok

Terence Dove

Ben Edwards

Roosa Frantsi

Mark Hales

Aleksandra Krajnc

Tuomo Kujala PhD

Esko Lehtonen PhD

Callum Mole PhD

Jussi Palomäki PhD

Jami Pekkanen PhD

Pasi Pölönen

Heini Saarimäki PhD

Juraj Simko PhD

Prof. Benjamin W. Tatler

Sanna Tiirikainen

Prof. Vincent Walsh

Prof. Richard M. Wilkie

Despite their efforts, many errors are bound to remain; these are of our own making and responsibility.

AND MORE PERSONAL THANKS

OL: Dad. For being sort of responsible for getting me interested in cars – and teaching me *very* early on that one must not always believe everything one reads in a book! Mum. For always being mum.

Prof. Heikki Summala, for giving me my first real crack at doing proper science, and for the wholly serendipitous (though perhaps not wholly unexpected) direction my whole career then went into! And to my TRU research group colleagues – most especially Jami and Esko – for all those hours spent on analyzing driver eye movements and vehicle telemetry... when the only person in the team genuinely obsessed with driving is just me.

And all my students past and present; for the honor and privilege of making me a teacher.

AD: Sorry mum that you didn't get the extension on the house you wanted because of all those years of karting we did, at least we've now got a book to put on the (bare) trophy shelf. Dad, of course; you sacrificed so much so I could get access to something only 0.01% of the world can ever do. I will never forget just how lucky I have been in my life to be involved in motorsport.

To my siblings Terence, Michelle and Doreen – can we turn our beds into bunk beds? It'll give us so much extra room for activities! Lewis, thanks for showing me what real speed was all about – Sonic!

My racing brothers. Usmaan, your friendship is more valuable than you know. Will, those conversations back in the day at 2am studying the art of kart driving made me the analysis-addicted person I am today. Oliver, thanks for reminding me to never play to the gallery.

SCIENTIFIC ENDNOTES

These end notes provide additional detail and references to some of the more technical scientific and driving technique concepts.

They are not necessary reading to understand and enjoy the rest of the book; they are here as a resource for those who want to go deeper into the scientific and/or race driving literature.

In researching for this book, I (OL) have read hundreds and hundreds of scientific articles, and dozens and dozens of books – I will only give selected references to those that I find particularly apposite and well-written; most of them are review papers from which the academically-minded reader will be able to follow up on the original research literature.

---

[1] We use the term 'racing driver' to designate all racers, whether their vehicle happens to have two, three or four wheels. (Some may even race snowmobiles or boats without wheels altogether).

Meaning that, for example, motorcycle racers are included in the designation of 'racing drivers'. Also whether it takes place on a track, on roads closed for a rally stage, or through the wilderness in desert raids racing is racing. So, rally driving is included, too:

Figure EN-1

A racing driver is a racing driver - no matter how many wheels they happen to be putting on the ground...

[2] This quote is from the book by Jenkinson (1959/1997) pp. 5-6. This work was a big source inspiration for this book. In a way our book is one long answer to the question: what can we say about that quality Moss is talking about, from the point of view of modern cognitive science and brain research? So, we begin the first chapter, in homage, with the account of Moss' and Jenkinson's Mile Miglia exploits.

On the other hand, we do not go to the fundamentals of driving technique in anything like the detail Jenkinson does – those have not changed and the original work is still remarkably fresh and current.

[3] Said Moss: '*the roads are closed – [but only] if people keep off them!*' In other words, as far as risk and uncertainty goes, in the Mille Miglia, there was the matter of whether or not all members of the crowd (or, say, stray dogs) would in fact remain at the roadside and not inadvertently wander into the path of a 150 mph missile with cross-ply tyres, and, by modern standards, not very good brakes at all... But these are matters that are beyond the driver's control... Or, well, not completely. Moss would deliberately 'snake' on straights, gently weaving from one side of the road to the other, in order to keep back the spectators! These were different times indeed...

[4] Jenkinson (1959/1997) p.43.

[5] https://www.youtube.com/watch?v=ft3VOCJnppg

[6] Sometimes the term 'expert' is used for anyone who has developed a skill to a higher level than the norm, or someone who has just some unusual skill or special knowledge but still at the level of everyday proficiency.

We will use the term 'expertise' as defined in the work of K.A. Ericsson, to refer to the ability to produce on demand world-class performance, including elite and super-elite, levels. This is the way the concept is used in the psychological research on expertise and expert performance. (Ericsson, 1993; Ericsson et al, 2018).

[7] John Fitch was Mercedes-Benz factory driver and an engineer; he would later go on to design, among other things, the Fitch barrier used on US motorways.

[8] Jenkinson (1959/1997). p.23.

[9] He is also widely considered one of the most versatile, implying an ability to develop a superb 'feel' for very different types of machinery (on the concept of feel see **Chapter 5**). On the variety of cars Moss raced, and his personal experiences of them, see Moss & Nye, 1987.

[10] On the concept of Situational Awareness see Endsley (1995).

[11] Jenkinson (1959/1997). pp.13-14.

[12] Note: When not talking about an actual person with established gender, we alternate the personal pronouns between the masculine and feminine.

[13] In contrast, look at a 1970s Formula One car. (Or may we suggest you point your browser to footage of a group B rally car – especially the rear-wheel drive Lancia 037, preferably with Mr. Alén at the wheel...). Here even the casual viewer can get a sense of the precarious balance between grip and skid where the expert racing driver lives, precariously balanced at a hair's width from disaster... You should Google up the video of Kenny Bräck in the wet in Adrian Newey's Ford GT40 at Goodwood...

[14] The term satisficing was introduced by Herbert Simon (1955). For an analysis of everyday driving as satisficing see Summala (2007), and Näätänen & Summala (1976).

[15] For references and discussion see the technical note on → *Synergies*

---

[16] **FLOW**

The concept of flow was first defined in its modern psychological form in the works of Mihaly Csikszentmihalyi and his co-workers (1975); see also Csikszentmihlayi (2014).

They characterized it as a state of *intense concentration*, *task focus*, a *merging of awareness and action*, and a *sense of mastery and control* one one's environment and actions, when one is totally immersed in a task, cut off from the rest of the world.

Subsequent research over the past 50 years has confirmed and refined the description of the phenomenological characteristics, and the conditions in which flow will occur. The commonly accepted conditions of flow are:

C1 task demands match capacity (skill and effort), at a high level of performance;

C2 the task presents clear and personally significant (proximal) goals; and

C3 the environment provides clear and unambiguous feedback on goal achievement.

These conditions are typical of situations where flow can be elicited.

The phenomenology of the flow experience is characterized by the following features:

P1 total focus in the task execution and concentration on what one is doing, to the exclusion of everything else in the environment (or one´s life);

P2 absence of reflective self-consciousness; merging of action and awareness; being 'one' with the task; disappearance of a 'self' separate from the 'doing';

P3 a sense of effortlessness and automaticity; a sense of not having to 'consciously intervene';

P4 a sense of being in control, confidence in one's skill and ability; absence of worrying about errors;

P5 the activity feels enjoyable or worth doing; the body feels good;

P6 a distortion of temporal experience (time may seem to go slower or faster than normal).

These are characteristics that people reliably report after they have been in flow.

Theoretically, Csikszentmihalyi maintains that flow is experienced, whenever all of attentional capacity (all of one's 'psychic energy') is concentrated on task performance. What this seems to imply is a unity of attention, awareness, or consciousness.

On the neural basis of flow see van der Linden et al. (2020).

---

[17] RACECAR ERGONOMICS

Many high-performance and race driving manuals begin with an explanation of how one should be seated in the car in order to properly feel and control the vehicle movements.

See e.g. Prost & Rousselot (1990) pp. 14-21; Senna (1993); pp. 10-23; Bentley (1998) pp. 11-14; Elford (2008) pp. 12-26.

On the design side of racecar ergonomics, Brown et al. (2020) give a brief history of racecar cockpit design. As to the specific layout of the twenty-or-so controls on a modern F1 car, they state that: '*examination of four recent F1 steering wheels revealed considerable differences in design philosophies indicating that teams may not be fully aware of the potential advantages of applying human factors (HF) methods*'.

In other words, they suggest that the ergonomics of the various mode and systems switches is still at a stage where different solutions are being explored, before converging to a best practice *de facto* standard.

---

[18] **STRESSORS AND AUTONOMIC RESPONSES**

Stressors the racing driver is subjected to include:

- physical effort of operating the controls (allometric contraction) and bracing one's body and neck against g-forces (isometric muscle strain)
- noise and vibration (from engine and tyres)
- shocks (from road surface irregularities)
- heat (ambient and heat soak from the engine)
- carbon monoxide (from exhaust gases)
- dehydration
- fatigue and not least
- the *mental* stress of concentration, competition and physical risk

See Watkins (2006), Kücükdurmaz (2012), Potkanowicz & Mendel (2013), Potkanowicz (2019) and Reid & Lightfoot (2019).

[19] The autonomic nervous system is responsible for many bodily functions like digestion, water balance, temperature regulation, release of hormones etc.

Eccrine sweat glands are found in the palms of the hands and soles of the feet. The biological purpose of these special sweat glands is not to regulate body temperature (like most sweat glands around the body), but to increase the grip of the hands and feet so you don't slip or drop things.

Of course, with racing boots and the gloves, the palms and soles are lined with grippy material – technology has augmented the properties of the skin, and so these evolutionary vestiges of our ancestral past no longer serve their original use for the racing driver.

---

[20] James (1890), p.402: "*My experience is what I agree to attend to*" (emphasis in the original).

---

[21] VISUALIZATION

Visualization is a widely used and studied phenomenon in sports, and there are multiple techniques. From a brain point of view it means 'driving' perceptual systems more top-down, based on memory, rather than bottom-up, on the sensory input.

For descriptions of the use of visualization techniques in racing, see e.g. Code (1983), pp. 30-31; Smith, 1996 pp. 3-4 ff and Bentley (1998) pp. 106ff.

For a review of mental imagery training in sports generally see Schuster et al. (2011) and Moran et al. (2013).

---

[22] UNCONSCIOUS INFERENCE

This is a technical term deriving from the 19th century physiologist and experimental psychologist Hermann vonHelmholtz to describe the fact that the way we experience the world in perception and consciousness is not *direct*, and we do not represent the world 'as the world really is'. Instead, our perception, memory, subjective experience etc. are in fact as the result of a complex and sophisticated process that combines the incoming sensory information to pre-existing knowledge (or assumptions – these may be learned from experience or built into innate, instinctive ways the brain works).

The end result of this process is to come up with its best guess as to what the world out there could be like. Because this proves is very fast and completely automatic, we are not at all aware of it – only its end-result. You can think of conscious awareness as 'the conclusions of the unconscious inferences'.

---

[23] THE RACING LINE

In simplified terms, taking a corner can be broken down into a logical sequence of successive steps - discrete phases identifiable relative to the geometry of the bend and the control actions taken by the driver to 'connect the dots'. (We discuss the cognitive basis as well as the limits of the 'connect the dots' metaphor in the **Chapters 3 -5**).

A concrete description and illustrations of 'the racing line' around a specific track can be found in e.g. Lopez, 1997 pp.103-125 and Holbert et al. (1982) pp. 89-97. From the point of view of vehicle telemetry signals see Segers, 2014, pp. 143ff.

The logic of how the racing line is put together as a kind of problem-solving exercise is described in Lopez et al. (1997) pp. 19-55. For a breakdown of the elements of a racetrack that will constrain the solution see Code (1983) pp. 2-9.

[24] For a highly readable and instructive physical analysis of the forces involved I would recommend Beikmann (2015).

[25] For a discussion of the too-early-turn-in-and-apex error, from a driving technique perspective see e.g. Code (1983) p. 23 and Bentley (1998) p.82.

The idea is as follows (referring to **Figure EN-3** below):

Driver A is eager to 'cut the corner' and turns in *too early* – perhaps under the idea he should '*drive as straight as possible*', or take a line that has '*the biggest possible radius*'... based on the corner's appearance. Or perhaps he is steering as a survival reaction to avoid the outside road edge rushing at him at high speed.

Whatever the cause, the consequence is he arrives at the apex too early. What is more, having taken an extremely large radius path into the bend, driver A has arrived at his early 'apex' with the car pointing in the wrong direction (the tangential line from driver A's false apex points too much to the outside of the bend) – and because he has been taking a wide line he has not had to slow the car so he is quite probably going too fast.

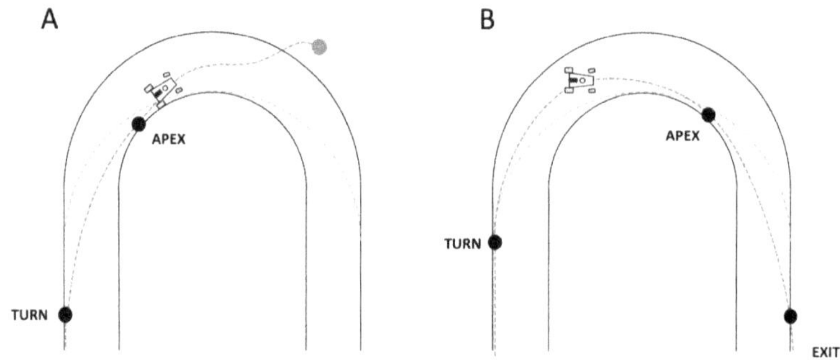

Figure EN-2

The dotted gray line shows the maximum *constant* radius path. Notice how the racing line for driver B is on the *outside* of the maximum constant radius path, and the line is also tighter, in the entry phase but *inside* the constant radius arc, and the line is wider, in the exit phase. (The opposite is true for the turning-in-too-early error).

Driver B, instead, is taking a proper racing line. Having chosen a late apex she is turning in later, then rotating the car more on entry – on a tighter line so that she can then get on the power and accelerate out of the bend smoothly, with a wide and widening arc.

This wider line in the exit phase means that the driver is not using up all the tyre's grip for cornering and so she can use some of the tyre's force-generating capacity to accelerate the car. The racing drivers' adage that sums up the above considerations is that '*slow in fast out beats fast in slow out (almost) every time*'.

(There is related to this a common misperception, however, that higher exit speed on the exit carries over to higher top speed on the following straight – but actually a slower exit loses you *time* on the straight, which you cannot make up… not *speed*, which you largely can. See analysis in Beikman (2015)).

This analysis shows how all lines that 'cut the corner' are emphatically *not* the same thing as the racing line.

Note that the turn-in point, apex and exit indicated in the figure are the three basic *waypoints* on a racing line.

See discussion on reference points in and waypoints **Chapter 3** and **Chapter 4**.

---

[26] See the end note on the → *Racing line* above.

[27] The concept of the friction circle is often attributed to Donohue & vanValkenburgh (1975). For a more technical discussion of the friction circle and the distinct (but sometimes conflated) g-g-plot (Holbert et al., 1982 p. 77) see Milliken & Milliken (1995).

It is fundamental to understanding the interrelationships between steering, accelerating and braking, and the effects of different vertical loads on the tyres.

---

[28] DYNAMIC LOAD TRANSFER

A less-dummy version follows. The distribution of *mass* of the vehicle does not change, as everything is strapped down and cannot move around. But how much of the total mass is suspended on each tyre – the distribution of *loads* on each individual tyre – can and does change dramatically. This is sometimes called dynamic *weight* distribution, but I prefer to use here the term load distribution because the term 'load' is probably, for most readers, less easily confused with mass than the term 'weight'.

As a rule of thumb, the total force – the combined total grip achieved by all four tyres together – is the higher the more even the loads are. The reason for this is the nonlinear dependency between load and grip. There are 'diminishing returns' in terms of grip from adding load to a tyre, yet everything you add to one tyre you are taking from another (with 'increasing loss'); therefore the tyre having load taken off loses more grip than the other that is receiving load tyre gains grip.Hence the notion of 'cornering flat' and 'balancing' the car at the limit of grip. For an explanation of how the weight (load) on each tyre and managing this dynamically affects race driving technique, see e.g. Dove (2016) pp. 92-109. (Cf. also Bentley (2003) p. 133).

---

[29] **STEERING**

The attentive reader may have noticed there are multiple distinct concepts of 'steering' at play here. Let's analyse them:

1. One concept is 'applying the correct amount of *steering wheel rotation.* This is further specified as amount required to...

1.1. control vehicle rotation to achieve a desired *path curvature,*

1.2. match vehicle *heading* with desired path *direction* and/or

1.3. keep the vehicle at a desired *lateral position* on a desired path

The correct amount of steering will change with speed, and depending on the understeer/oversteer characteristics of the vehicle; an understeering vehicle will need more lock as speed goes up, an oversteering vehicle less.

2. The other meaning of steering is 'controlling vehicle trajectory (i.e. lateral position and orientation) relative to the road or a desired path... *by whatever control inputs necessary* (i.e. just the steering wheel).

When the vehicle dynamics are very simple (in everyday driving where load shifting between tyres is minimal and tyres are in a linear response regime), these two meanings amount to more or less the same thing – but when vehicle dynamics are more complex they do not (in a racing car at the limit of grip where tyre response is nonlinear and there is substantial shift of load across tyres). In science, the simplification is often made in theoretical discussions, driver models and driving simulator experiments (especially so in psychology) where *the steering wheel* is considered as the means to effect *lateral control* and the *throttle* and *brake pedals* as the means to effect *longitudinal control.* This does not hold for controlling a race car or a racing motor bike at the limit, because of the strong dynamic coupling between the controllers.

Understeer likewise has a number or related but distinct meanings, and racing drivers, engineers and psychologists use the term somewhat differently:

1. To a racing driver, understeer means the car does not 'turn as much as intended' (in the extreme case: front wheel lock-up or skid where directional control is completely lost. Oversteer is the opposite. In the extreme case: rear wheel skid where the rear axis does not steer at all).

2. To an engineer, a vehicle understeers when an increase in speed requires more steering wheel lock / front tyres assume larger slip angles than the rear. (Oversteer is the opposite).

3. A psychologist says a driver understeers or oversteers if they steer so much that they end up taking too wide an arc or turning too tightly relative to the path curvature of a reference path (i.e. they drift wide or cut inside a reference path. Note that 'over-steering' in this sense, cutting the corner, is in fact the normal strategy in both everyday and race driving!

Oversteer can mean many things: slip, skid, drift and more,,, These are concepts that are often and easily confused.

Turning the steering wheel creates a *slip angle* between the plane of rotation of a tyre and its direction of travel.

Note: I will call the direction of travel *heading*, although in vehicle engineering this would be referred so as bearing – for historical reasons engineers and psychologist tend to use the term heading differently – as in '*which way are you heading?*' even at the risk of some confusion, I will follow the terminology in the experimental psychology of steering.

Now, only a tyre with a slip angle generates a centripetal (sideways, lateral, steering) force. At the same time it inevitably creates drag (see figure below) Thus, steering the car changes speed. Thus, rotating the car by turning the steering wheel inevitably affects its speed.

Note that slip angle must not be confused with steering angle, which is the angle between the plane of rotation of a tyre and the vehicle longitudinal axis. These do not have the same value – indeed, they can be in opposite directions as in the figure below where the driver has applied opposite lock.

Referring to **figure EN-3** on the next page: although the wheels are turned to the *left*, the car is in a *right-hand* turn – think of the direction of travel as rotating clockwise – it's just that the car is a bit sideways. The front tyre is nevertheless still slipping to the *right*, which means it is steering to the *right*.

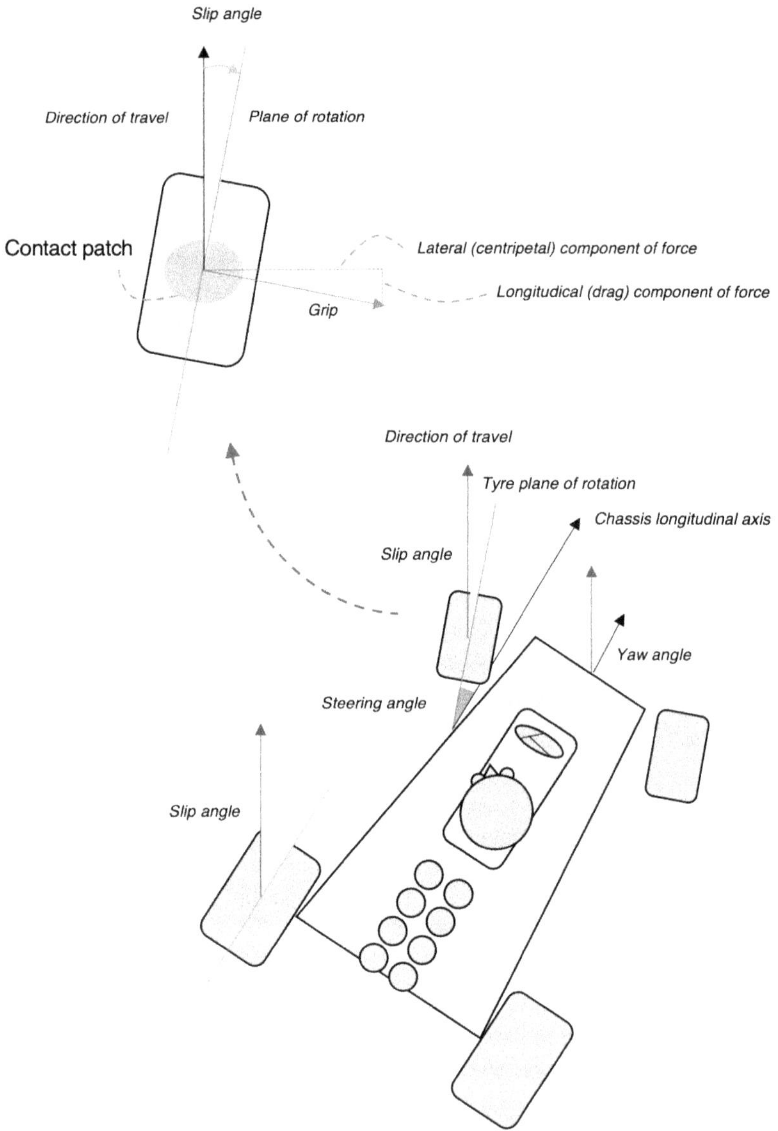

Figure EN-3

Notice how the rear tyre's slip angle is much larger than the front tyre's slip angle. This is an oversteering condition. And as long as the rear tyre is not skidding it Is also steering.

(Rear wheel steering angle is not shown, but the plane of rotation will generally not be *exactly* parallel to the chassis.For example, the outside rear tyre might here assume some toe-out – this is engineered into the suspension, even in cars without 'rear-wheel steering'.)

On the other hand, changing the speed (longitudinal and lateral acceleration, using brakes and the throttle) move loads front-to-aft, changing the force a tyre produces (at the current slip angle), thereby inevitably affecting the direction of travel. Speed and steering are coupled, and both are controlled by all of the controllers (steering wheel, brake pedal, throttle pedal). Similarly (only) a tyre that rotates faster than a rate that matches its speed of travel generates a longitudinal force to accelerate the vehicle – the tyre in this condition is said to have *longitudinal slip*.

Here is an important point: *slip*, must be separated from a *skid* which is a condition where the tyre is sliding on the surface. A skidding tyre cannot produce any sideways force, as all of the grip is 'used up' longitudinally (for braking, like in a lock-up, of accelerating, in a burn-out, or if the slip angle is too large, as in an 'oversteer moment'). Slip is *necessary* for a tyre to steer, skid *prevents* a tyre from steering.

So technically, in 'engineerese', a rear wheel skid is not at all the same thing as oversteer – even though in racing driver jargon it often is used synonymously. The rear tyre above, for example, need not be skidding. I.e. it may still contribute to steering in a four-wheel-drift. And, if that was not enough, then there is also yaw slip or side slip, which refers to the difference between the vehicle's direction of motion and the direction where its nose is pointing, i.e. its heading. When it comes to vehicle dynamics there is hardly a technical term more slippery than slip ...except...maybe *heading*... Because the way I just used the term 'heading' is in the way it is used in automotive and aerospace engineering and nautical terminology... but elsewhere in the book it is actually used in the way it is used by psychologists, where it refers to the *instantaneous direction of travel!*

The reason psychologists use the term differently is historical: in the 80s and 90s they were interested in how well humans could estimate the simulated direction of motion from various types of animated displays in the presence of side slip. Like, when you are racing on a sim and you are in a skid, how well can your brain estimate 'what direction the car is heading.' This became to be called 'heading estimation'. (If the term 'heading' had been used in the engineering sense, the task would be trivial: an engineer's 'heading' is after all always at the centre of the display! Assuming there is no head yaw angle rotation simulation where the virtual camera rotates in the cockpit, that is...). Working in the field of driver behaviour and modelling you need to wrap your head around a pretty complicated mechanics of different coordinate systems... and on top of that different fields referring to the things in different words.

Phew...And don't even get me started on 'feedback' and 'feedforward' control!

Well, there *is* an end note on → *Feedback and Feedforward Control...*

---

[30] There are a number of slightly different versions of this famous quote, and I have not been able to trace down where it was originally from. Mr. Andretti may of course have said it on *numerous* occasions!

---

[31] MODELLING CHOICES

How one chooses to analyse and classify things is what scientists and engineers sometimes call a *modelling choice.* And the first steps you make here have a tendency to get locked in. They become resistant to change. So, it makes a difference that you set off on the right foot or, seen the other way around, the choice of framework could be done in a number of ways, depending on the picture you eventually want to build up to.

This is all a bit philosophical perhaps, but it makes a real-world difference: once your mind gets into the habit of categorizing things from a particular viewpoint, then that viewpoint becomes your fixed frame of reference and you tend to relate *everything* else to it. Soon it will no longer be questioned, unless explicitly held up and contrasted with other ways of looking at things. (And this type of thinking is hard to do!)

---

[32] ILLUSION OF UNDERSTANDING

There is a mental pitfall here to avoid. Unless you go through the trouble of looking at the task of driving with some rigor depending on your background you could end up just mapping unfamiliar, and meaningless names of cognitive processes and brain parts to a loose mental picture of 'what driving feels like', or mapping unfamiliar and meaningless names of driving techniques to how you think the brain works (if you are a neuroscientist)... or mapping meaningless names of driving techniques to meaningless names of brain parts!

This would produce at best confusion or, worse, an *illusion of understanding.* The term *illusion of understanding* (Wilson & Keil, 2003) refers to the finding that people are not very good at all in estimating how well they understand how a system works:

'*When asked how a helicopter works, they seem to think they have knowledge approximating a detailed annotated drawing, but actually have a much coarser understanding corresponding to little more than the sense of a thing with blades that turn and provide lift.'*

This illusion is specific to the kind of *explicit explanatory knowledge* that involves being able to describe accurately complex networks of causal and logical connections.

When people estimate how deep their knowledge of *facts*, or how apt their *procedural* knowledge or how well they understand information that is in *narrative* form, they are much better at it.

---

[33] Three rotational degrees of freedom for the head and each eye, plus one for steering, brake, throttle, clutch and gear selection each. This relatively low dimensionality of the system alleviates the so-called curse of dimensionality for modelling vehicle control – a very useful characteristic of motor sport expertise for mathematical exact modelling of the skill.

Yet even here there is as much room for developing a skill, and a deep understanding, and room to express individual style and flair in the particular solutions one finds to the problem of going faster.

[34] For a historical and conceptual overview of computational steering models in engineering and psychology, see Lappi & Mole (2018). For a review of the psychological models (and how they relate to gaze strategies) see Lappi (2014). For technical reviews of the engineering models see Macadam (2003), Sharp (2011), Nash et al. (2016).

[35] Of course to be specific you always have to specify what *frame of reference* 'lateral' and 'longitudinal' are meant to be taken relative to.

Common choices are vehicle direction of motion, vehicle chassis longitudinal axis, tyre direction of motion, tyre plane of rotation etc. For the level of detail in most of the book, this level of exactness is not necessary.

[36] On the different meanings of 'steering' see the end note on → *Steering*

[37] In general, you can always look at causes and effects in either of two ways:

cause → effect ('if I do this what is the outcome?') or

effect → cause ('if I want this to happen what should I do?')

The former is called *forward inference*, and the latter *inverse inference*. See also the end note → *Unconscious inference*

[38] There is another important way the connect-the-dots analogy breaks down, (related to the discussion on car-control in **Chapter 5**).

In a child's connect-the-dots picture – the analogy here – the 'waypoints' are already, physically there. They have fixed, immutable locations on the page. They are designated by visual markers and a number beside it that tells you where it sits in the sequence. This is obviously not the case on the racetrack.

But there is a deeper, subtler, point as well, which I often find people miss when I first tell them about the waypoint hypothesis. Real waypoints are dynamic. It's not just that there is no physical dot marked on the track to tell you where you brake, where you brake is not always exactly at the same place.

Each action defining the waypoints will change with variations in speed, balance; moving from one phase to the next *cannot* be perfectly pre-programmed and fixed in the driver's mind/brain, in advance. The racing line emerges and unfolds over time. For example, how you get on the throttle depends on how you apex and where; and apexing itself has to be adapted to how the entry was taken etc. This is as if where and how the first line is drawn in the picture (where and how you get on the brakes) would actually shift all the dots in the picture slightly – where exactly the next dot is, and thereby what line you should draw to get there adapts. (In racing this adaptation of course happens completely inside the mind).

And remember: the *control actions* will have to correctly match all the different *controllers* – thus the *turn-in point* is not about only *where* and by how much you turn the *steering wheel.* It's about, how you are getting off the *brakes* at that point as well, because this will have an effect on dynamic load transfer. And in the motor program, at the control level, it's also about *when* you turn the wheel and release the brakes – not just about the 'where' of the navigation level.

Cf. also end notes on → *Waypoints* and → *Timing*

[39] Physically, the distance between the braking point and the turn-in point and the difference between the speed from which one starts decelerating and desired entry speed determine the amount of speed that needs to be shed, the required deceleration, and the actually required braking distance.

[40] The heel-and-toe sequence is shown e.g. in Senna (1993) pp. 16-17 and the double declutching sequence is depicted in Lopez et al. (1997) pp. 95-96.

[41] See the end note on → *Dynamic Load Transfer*

[42] The 'classical' technique is to be off the brakes at turn-in, but a part of advanced corner entry technique is trailing off the brakes or *trail braking* in a way that accurately adjusts speed as well as front-rear load to distribution all the way from the turn-in point to the throttle-on point.

[43] The terms slip, skid, drift, oversteer are often used interchangeable when talking about driving technique and car behaviour.

See e.g. Milliken & Milliken (1995) for their accurate engineering definition, and also the comments on oversteer in → *Steering*

[44] *Steering wheel* angle may not be at all constant but subtly manipulated, jointly with the throttle, to control the vehicle direction (as well as the yaw slip angle and tyre loads). The driver is using the steering wheel not so much to change path radius – not so much to 'steer' that is! – but is rather 'balancing' the car with the steering wheel and the throttle.

Indeed, at this point they will more likely use the throttle to steer the car, inducing some oversteer or understeer by shifting load between the front and rear axle. The actual value of steering wheel angle during this stable state will depend on the yaw slip angle of the vehicle (yaw slip refers to the angle between the vehicle's direction of travel, and the direction its chassis nose is pointing).

Cf. the different meanings of → *Steering*

[45] So, by unwinding 'steering' we here really mean *unwinding the trajectory,* thereby 'steering less' – trading off path curvature for speed.

There are different ways this can happen that of course involve the use of *both* the steering wheel and the throttle.

[46] At a higher level, factors such as fuel and tyre saving can be critical for race success. We are only talking about a single bend here.

[47] McRuer et al. (1977) hierarchy was also the basis for the two-level framework of Donges (1978), a central framework theory in vision science and traffic psychology.

See Lappi & Mole, 2018 for a thorough historical and conceptual analysis), and hierarchical organization of the driving task is a standard approach in traffic psychology.

See Lappi (2018) for an analysis of some race driver training techniques in the McRuer framework.

[48] Many textbooks spend a lot of time explaining reasoning that goes into this.

For example see the break-down of a lap of Sebring in Lopez (1997).

[49] Code (1983), p. 19.

[50] This story was recounted to Alan by Alex Hawkridge in a podcast interview and is also found in Hilton (2010) pp. 154-155.

[51] There is an element of conventionality in what gets designated as one bend. The 'straights' between bends might not be completely straight (like the run up from Ste. Devote or the Tunnel at Monaco), and a bend might have multiple radii (like Massenet or the Chicane), so for some purposes these could be considered two distinct bends as well. For successful *communication* what matters is that everybody agrees to where the bends are, and their naming or numbering – not that you get it 'right" (This applies to many other aspects of verbal communication as well).

[52] The key here is the rate of information the track throws at you. The rate of information in the automotive context is discussed and analysed in Senders et al. (1967) and Kujala et al. (2015). This is also discussed in Chapter 8.

[53] Note that the way we talk about it here, we are perhaps making it sound like the knowledge of the driver as if the driver's head contained 3D scale models, of finer and finer resolutions. Obviously, this is not literally so.

We are describing really the track features and scene structures out there. They must – in one way or another – be 'known' to the racer's brain, but the form this knowledge takes is not going to be an accurate 3D replica

This is sometimes called the knowledge level. Though even here one must be a bit careful. The actual content or the form in which the information is represented in the brain is as yet unknown.

We can tentatively describe the knowledge by using words to describe the object known – 'as if' the knowledge were represented in a 1:1 scale model. Which everybody knows it isn't. The form of the representation of that knowledge might be chunks, prototypes, schemas, concepts, neural networks, or whatever. And even what a particular kind of system actually needs to know is not always as clear as these examples can make it seem.

[54] The exact number of bends on the Nordschleife is a matter of interpretation:

http://www.greenhellguides.com/2014/08/24/many-curves-nurburgring-nordschleife/

---

[55] **WAYPOINTS**

The term waypoint may be a bit unusual here – it's part of the waypoint chunking theory, not part of established racer jargon.

So what do I mean by calling the braking point a 'waypoint'? It simply means a specific location where the braking action happens.

It need not be exactly at any visible reference point, and it need not be at the same place every lap. But what it is is location that you can point to when you are observing the racing driver. In Lappi (2014) I define waypoints as '*any point on the visible path, such as a marking on the pavement, or, if the driver plans his FP* [future path], *a point such as the turn-in point (where the driver will turn the wheel entering the bend) … they need not in principle present any distinctive visual feature.*'

---

[56] EXPERTISE AND EXPERT MEMORY

For a good general introduction to the scientific study of expertise in chess and beyond, see Ericsson & Pool (2018).

[57] Although, based on surface appearances, the analogies between racing and chess may be surprising to common sense, nevertheless from a cognitive theory point of view it would actually be rather surprising if it *didn't* hold. (Because race driving would in that case be somehow quite different from virtually all other types of human expertise studied).

[58] Just having a huge knowledge base in itself is not enough. As the amount of stored information increases, this also presents a problem in itself – one you may be familiar, the more so the bigger your garage: *How do you find the right piece of information* (or right tool) *quickly and efficiently enough for effective use*? This is one aspect of what in Artificial Intelligence is called the Frame Problem (see Dennett, 1984, for a very readable explanation).

Human memory seems to be really good at this, in ways that give human intelligence a flexibility and generality that Artificial Intelligence has not been able to crack.

An entertaining account of expert knowledge, how it is organized, its use in inference etc – is illustrated by the fictional character of Sherlock Holmes – is given in Didierjean & Gobet (2008).

[59] It is an interesting question whether timing – down to how and when you pick up and put down the physical pieces – the rhythm of your play is important for some players.

(The way it is, for example, in snooker. It is a more self-paced game than race driving where the speed you travel dictates how hast you need to decide). The question of timing the choice of moves at different temporal scales is an interesting and to my knowledge very little studied aspect of expert performance.

Cf. also the end note on → *Timing*

[60] If you think this comparison made in jest, ask yourself whether you might be prejudiced. You have to be *good at thinking* to be a great champion, I believe; and JYS has a clarity of thought and expression I genuinely admire.

[61] Classical studies on chess expertise: DeGroot (1946/1964), de Groot & Gobet (1996), Chase & Simon (1973); Simon (1974); Gobet & Simon (1996a-c, 1998). For review see Gobet et al. (2001).

---

[62] With the rather obvious exception of the opening position, which is always the same, and a limited number of best opening games which are repeated, because it has been worked out what are the most useful developments and countermoves. Therefore the discussion applies to mid-game and end-game positions.

---

[63] MEANINGFUL PATTERNS

It is important that the patterns chess experts encounter in real games have been generated by players – by a selective process of only making *meaningful* moves. Experts are not superior in 'reading' *meaningless* patterns like chess pieces randomly placed on a board.

Real game positions have been generated by the constraints of the rules of chess on the one hand, and on the other hand players actively seeking to create situations – attack, defense, power bases etc. – that would give them an advantage.

Similarly, real race tracks are generated by the laws of physics and geometry that the track engineers work with, to create interesting or meaningful track configurations. Over time an expert will have been exposed to, and will have become familiar with, *the subset of meaningful patterns.*

For example, for a chess player, whether a pawn is placed in front of your king means something in relation to the rest of the pawn structure; and the way it affords defensive moves or blocks the opponent's attacking moves, depending on where exactly it is positioned. Similarly, a visual reference associated with the start of the braking area or the curve apex will only mean something to the racing driver in the context of the arrangement of other references and the 'moves' you can make or not make on turn entry and turn exit.

Depending on where exactly you choose to begin to brake, or how early or late you decide to cut in to the apex, will create different lines. So: where, metaphorically, 'the brake marker or the apex marker is placed' becomes part of a pattern of interrelated reference points, whose placement is associated with different 'lines of play'.

If you go to a driver training session on a race track, the instructors may place *actual physical cones* at the suggested braking point, apex and exit, to give you visual reference points. This allowing you to quite literally 'connect the dots', driving from one waypoint to another. They make explicitly visible some of the relevant, or meaningful, physical relations. Just like drawing arrows, circles or other graphical cues into chess diagrams!

However, for the expert the waypoints are purely mental, and the 'chunks' and 'lines of play' are not out there, in the optical information. It is largely 'in the eye of the beholder', i.e. in the mind.

---

[64] Glaser (1985).

[65] General intelligence refers to the things you are tested for in standardized IQ tests. E.g. if you want to join Mensa. Generally speaking, these abilities are not the same thing as the skills you use to excel in your field of expertise.

[66] To what extent differences in expert performance depends on special talent for particular domains, or domain general cognitive ability a.k.a. general intelligence, or lack thereof, remains contentious.

That is, while it is generally accepted a large amount of experience is *necessary* for expertise, opinion of researchers remains split as to how *sufficient* it is.

---

[67] **PROCEDURAL KNOWLEDGE**

The psychology of learning and the human memory system makes a fundamental distinction between *procedural knowledge*, *knowing-how*, and *declarative knowledge*, *knowing-that*.

Declarative knowledge is knowledge that can in principle be expressed in words and communicated, such as what you had for breakfast, what the capital of France is or what you had for breakfast last Tuesday. If an individual can in fact recall and express their declarative knowledge it is said to be *explicit*, if the information is stored but cannot be at the moment recalled it is *implicit*.

Procedural knowledge for the most part cannot be expressed in words – it can only be expressed in action. It therefore mostly implicit: we cannot say what we know.

Procedural knowledge resides in the brain in largely distinct subsystems to those of declarative memory and language. (as discussed in **Chapter 8**).

---

[68] See H. Dreyfus (1992) book *What Computers Still Can't Do.* It is a sequel to his earlier book *What Computers Can't Do*, and given the pace of development of Ai in the past 30 years and the past 10 years in particular, writing a book with such a title risks quick obsolescence... yet the general point stands that whereas AI systems achieve human-level or even super-human performance in tasks such as Go (Silver et al., 2018) and computer games (Mnih et al., 2015), including the real-time strategy game StarCraft2 (Alphastar Team, 2019) and even sim racing (Fuchs et al., 2020; Remonda et al., 2021) – the current systems go about it in *fundamentally different ways* than the human mind.

For an explanation of Deep Reinforcement Learning AI playing Gran Turismo see: https://www.youtube.com/watch?v=Zeyv1bN9v4A

[69] It's in the learning phase that a process of deliberate studying of alternative openings and endgames that search takes place. Learning is searching, learning is thinking – searching new paths in the space of 'lines of play'

In chess this is nowadays done with the help of a chess computer. For a jazz artist, a downhill skier or, indeed, a racing driver this search process is perhaps less regimented, and less reliant on explicitly, consciously thinking about all the 'angles' and alternatives. But the underlying cognitive dynamics of proceduralization of the search results are probably similar.

[70] This process-based rather than outcome-based approach to the nature of expertise is expounded in Bereiter & Scardamalia (1993).

This way,, to define true expertise there is no need for some arbitrary fixed amount of experience or level of skill. E.g. a 10 000 hour rule.

[71] Interviewed on Autosport.com Top 50 drivers of 2018 feature by Edd Straw.

[72] About working memory capacity, see e.g. Engle (2002) and Conway (2003). Chunking allows experts in a domain to bypass some of the limits of WM capacity.

[73] See Ericsson et al. (1993), Ericsson (2020), Ericsson & Pool (2017), and the 'ten year rule' popularized in Gladwell (2008).

[74] Bentley (2003) p.50

[75] For overview see e.g. Glaser, 1985; Gobet et al., 2001.

[76] Loomis et al. (2013). See also Tatler & Land (2011) and discussion in Lappi (2016).

[77] ALLOCENTRIC AND EGOCENTRIC FRAMES OF REFERENCE IN DRIVER MODELLING

That the navigation level wayfinding information – reference points and waypoints – are fixed in *allocentric* space means that their known relative directions and distances do not change as the car moves on the track. Their *egocentric* directions and distances relative to the driver change, but in an allocentric frame of reference all these changes can be represented as movement of one object (the observer). In contrast to waypoints, travel points (like virtual targets ahead of the car on the ideal line, such as the ideal line help or a ghost car in many racing games) move in the world so that even as you steer by them you *do not intercept* them.

If the chunking-based analysis is on the right track, it could point to a new sort of approach for driver modelling as well. One that is derived from general properties of human memory as understood in expertise research, and analysis of the task structure at the navigational level. It should be mentioned that this kind of cognitive-level analogy between reference points and chess pieces is radically different approach from current computational models in vehicle systems engineering and psychology that describe steering control. By and large, the scientific study of racing driving has so far largely ignored the navigation level. In vehicle dynamics research, for example, sophisticated driver-in-the loop simulation models have been developed, but these models concentrate mostly on the control level (which we will cover later), and we should make clear that there is no exact, mathematically rigorous definition of what a reference-point chunk would be.

Most current steering models represent the information extracted from scene preview as either a desired path, or one or more steering points that are purely egocentric *travel points* – i.e. points that move along a desired path in 3D space, which the driver model then follows.

(Rather like a line assistance feature in some racing games, where a kind of visual tongue, identifying the ideal line, protrudes from the front of the vehicle).

In these kinds of models, the navigation and guidance level are simply assumed to provide, to the lower level being modelled, one or several target points in the visual field. They are tracing out the perfect line in front of the vehicle out there, and assumed to be 'observable' to the model.

(The difference between a travel point and a waypoint and its implications for steering control are explained more technically in Lappi (2014) and Lappi & Mole (2018)).

This reduces the problem of steering into a more simple task: chasing a moving target that is, rather magically, moving on optimal path, and can be observed even though, visually, there is of course nothing there. Consequently, such models are silent on the trajectory planning (chunking the visual scene) needed to determine where the desired driving line is. The picture of static reference points and relations that chunk them, however fits rather well with descriptions of how one should go about learning and thinking about cornering that are found in the racing driver training literature.

---

[78] Moss & Nye (1987) p. 63.

[79] Chater (2018) argues this is how conscious thought mostly works: making up a coherent story of things automatic processes in memory and perception throw up. Our self-reflection is therefore a quite an unreliable indicator to what information we *actually* use to make decisions.

[80] What psychologists call *situated* knowledge – knowledge embedded in, and only efficiently retrievable from memory in the physical situation of its use.

[81] Csikszentmihalyi (1975/2000, 2014).

[82] Bentley (1998) p. 105

[83] Crundall & Crundall (2012).

---

[84] LOOKING AHEAD AS EXPERT DRIVING INSTRUCTION

Actually, it is not at all obvious why looking further in *space* will give you more *time* to react – given that whether you look below a thing in the visual field (closer than an object or location) or above a thing (further than an object or location), how far in the visual periphery the thing is the same. That is, the visual angle between the direction of the object and the direction of gaze is the same. This suggests it is not the position in the visual field (in what *direction* things are relative to gaze) that is the primary reason, but the actual *distance* in depth.

(Although it has been established that there is an asymmetry: the *lower* visual hemifield has a finer-grained attentional resolution than the upper visual hemifield. Attentional resolution by the way is not the same thing as visual resolution – it is much more coarse).

Physiologically, it might have to do specifically with the focusing of the lens, stereo vision and/or the way the brain processes depth and/or the way the brain processes *actions* (the act of 'pointing' gaze far ahead might turn out in itself to facilitate other motor actions oriented to far distance, regardless of the actual change in visual input or where on the retina the different objects and locations end up projected to!). This is a fascinating and to my knowledge very little studied question.

[85] According to the premotor theory of attention (Rizzolatti et al., 1987; Rizzolatti & Craighero, 1998) a shift of attention simply is preparing (but not necessarily executing) an eye movement. And eye movements are not just about direction – your brain programs eye movements to point gaze in a particular direction *and* at a particular depth… might this be the neurological reason behind the 'look far enough ahead' instruction?

[86] The alternative remedy would be '*steer and you will automatically look*': develop a habit to avoid the obstacle, and your eyes, and memory recollection, will automatically follow. At the kinds of (sub-second) timescales we are talking about here, it is actually difficult to tell these apart. At such fine granularity there may not be a neat, linear, sequence in time where attending, looking and seeing could be ordered in a specific linear sequence. (Even when commonsense intuition may say that there surely *must* be). They may run in parallel and overlap – if they can be disentangled at all.

---

[87] The part of the brain constructing the narrative – sometimes called the interpreter (Gazzaniga, 2013) – may have surprisingly little to do with the parts of the brain actually making the decisions.

For a great introduction to this line of research see Chater (2018).

[88] See e.g. Chater (2018). Memory for events can be an unreliable guide to the actual mechanisms – especially flash*bulb memories* for novel, threatening situations or 'important events' that are vivid and persuasive, but can in fact be highly reconstructive.

(See Hirst & Phelps, 2016, for a critical overview of the flashbulb memory literature)

This may be partly because they get consolidated by the amygdala not the hippocampal system – part of the subcortical, emotional, automatic limbic circuitry of the brain, rather than the explicit, deliberate, conscious systems.

[89] Bentley (1998) p. 105.

[90] In experimental psychology, directing attention to a location of the visual field but without the gaze moving there is called directing attention *covertly* (Posner & Cohen, 1984; Hunt & Kingstone, 2003).

[91] Burr & Morrone (2012).

[92] There are a few research papers on racing drivers' eye movements (Land & Tatler, 2001 and Lima et al., 2019, in the field and vanLeeuwen et al., 2017 in a

simulator). The eye movement behaviour of a racing driver is different from the extensive visual scanning on the road.

[93] Code (1983), p.32

[94] And during sleep too, in the Rapid Eye Movement i.e. REM phase of sleep.

[95] Visual scanning in natural tasks more generally is described in Land (2006). A good source to understand peripheral vision in driving is Wolfe et al. (2017).

[96] You can get a good idea of how an expert driver (a driving school instructor) scans the road from Lappi (2017), and especially the accompanying videos:

https://figshare.com/articles/media/Supplementary_Movie_1_full_video_mp4/4498466

https://figshare.com/articles/media/Supplementary_Movie_2_17_21_31_r_mp4/4498613

https://figshare.com/articles/media/Supplementary_Movie_3_17_23_48_l_mp4/4498619

[97] Code (1983) p.28

[98] Widescreen visual attention and drills related to it are described in Code (1983) pp. 32-33; Code, 1993, pp.92-94; Bentley, 1998, p.105; Bentley, 2003, p.50,

[99] Bentley (2003) p.50

[100] If we hypothesize that the reference points are known locations in an (allocentric?) cognitive map, then the choice of where to look, or 'adjusting RPs' means that there must be RPs stored in long-term memory that are associated with current visual scene.

One hypothesis is that they are navigational information *chunked* into long-term term memory (see the end of the **Chapter 3** on Navigation).

[101] Code (1983) p.19

[102] See e.g Bentley (1998, 2003).

[103] Bentley (2003), pp. 19-20.

[104] It may be *technically* noise in a mathematical model, in the sense that it is variability that the model cannot explain. But of course, a model is a simplification. (And so: *for a more complete model the same variability might not be 'noise'*).

[105] Textbooks – besides giving you mathematical formulas to plug in numbers for the various forces and locations – will illustrate the ideas by geometrical depictions of the physics of the situation.

Diagrams, such as free-body diagrams, that help you quite literally to see the physics through an *analog model*: the relations among the points, lines and arrows recreate the relations between body parts.

There is a reason for this. Language is not very effective for communicating geometrical information (natural language that is – artificial algebraic languages used in mathematics and physics are different, they are designed precisely for expressing such things).

[106] Randy Beikmann's *Physics for Gearheads* (Beikmann, 2015) is highly recommended if you are interested in this.

For more serious interest, Milliken & Milliken's *Race Car Vehicle Dynamics* (Milliken & Milliken, 1995) and Jörge Segers' (2014) *Analysis Techniques for Racecar Data Acquisition* cover all that you are likely to need to know, unless you actually pursue a degree in automotive engineering or driver performance.

[107] The body schema is also based around control: 'you' is what you can directly control. What you cannot directly control is 'other'.

But as control over a tool, such as a hammer increases with experience – or, plausibly control a vehicle such as a race car, although nobody has yet done the detailed neuroscience to look into the matter empirically – then the brain begins to treat it as a part of the body (the technical term neuroscientist use is to say that the tool is said to be appropriated to the *body schema*). That is: the parts of the brain that monitor and control events in your body and keep track of where the boundaries of the body lie, begin to represent the tool as part of the body!

[108] This unification came about in 1940s with the mathematical description of servomechanisms using block diagrams, and the theory of cybernetics. It was extended into modern feedback and feedforward nonlinear systems control, filtering and state space analysis in the 1960s and 70s. Since then, it has become a huge and highly technical field, used extensively in e.g. autonomous robotics and AI.

A technical introduction to basic concepts of control theory for non-engineers can be found in Jagacinski & Flach (2002).

[109] A good history of dynamical systems theory is Gleick (1987).

The analysis of coupled dynamical systems requires an overall view – qualitative changes in behaviour often cannot be traced back into a single even in (or outside) of the system, that could be isolated as 'the' cause. In a closed-loop system, where the output of the dynamic system can feed back to influence the system, the system can have an effect on itself in 'circular causality'.

[110] The system's behaviour could be described in mechanical or in informational or energetic terms. It is simply taken to be a dynamical system consisting of observable variables and controllable variables.

[111] I am obviously making huge simplifications here – which is the point of *modelling.* A model takes a system that is too complex to understand and represents some aspects of it with a model that is simple enough to understand. A computational model, moreover, allows model behaviour to be investigated by mathematical analysis and numerical simulation, that allow a more precise understanding of the model than just a conceptual or intuitive understanding.

[112] This is loosely put in the main text.

More specifically: if the controlled variable itself is not directly fed back to the controller, the control is technically open loop. Other variables in the output might be fed back, making the control apparently feedforward control – but only in the semantic sense that it depends on what one has chosen to designate as the controlled variable.

This distinction arises concretely in visual steering models, when 'the' controlled variable is considered to be the lateral road position, and the (feedback) signal used by the controller is preview of a steering point further ahead on the road.

Yet the control is pure feedforward control, because turning the steering wheel not only changes the lateral road position, of course, but also rotates the car. So, calling this kind of online, model free, use of visual preview 'feedforward' or 'anticipation' is somewhat misleading. It is more properly called prospective control.

For more detail on the models themselves, and for a more technical theoretical discussion see Lappi & Mole (2018), and also the end note on → *Feedback and feedforward control*

[113] Technical jargon note: this is called Proportional (P) control.

Control can also be proportional to the rate of change of error, called derivative or difference (D) control, or to the integral or sum of error over a period of time called integral (I) control. A feedback controller can combine all three in PID control.

I.e. what the text refers to as feedback control is PID control of observable output variables.

Using feedback in model-based control is here called *predictive* control. This should be understood as opposed to *prospective* control, which is being considered here as an *anticipatory* form of *feedback control.*

See end notes → *Feedback and Feedforward Control* and → *Predictive control*

[114] While physically acceleration is just the change of velocity, and the brain might 'infer' acceleration by comparing successive observations of speed, physiologically acceleration can be sensed more directly and accurately by sense-of-balance organs in the inner ear.

These organs cannot sense the value of constant speed, however – this signal source needs to be integrated to give an estimate of speed and integrating twice gives estimate of location. These estimates can be combined with e.g. visual estimates in multisensory integration.

[115] High anorak-level alert. When Porsche moved from air-cooling to water-cooling for the 986/996 series, the sensor for the feedback signal is at the *radiator* end of the circulation, not at the *engine* end, giving M96 engine owners some anxiety (the Boxster/Carrera range – GT3 and Turbo models used a different engine). This is exacerbated by the absence of an oil temperature gauge, and the fact that the coolant temperature gauge readings are not calibrated to degrees Celsius. Ok, anorak levels returning to normal.

---

[116] FEEDBACK AND FEEDFORWARD CONTROL

There is some confusion, even in the technical literature, in the use of the terms 'feedforward', 'anticipatory' and 'predictive'.

Many people are using them interchangeably as synonymous, which fails to take into consideration the fundamental difference between *prospective* and *predictive* anticipation. (Prospective anticipation is not genuinely feedforward).

In model-free prospective control the control is open-loop from the point of view of the controlled variable and can anticipate changes in the controlled variables – but it is nevertheless a closed-loop reaction to *observed* change in *some* variables. It is not based on a prediction of any variable.

PID responses to *observed* values of preview variables that are *ecologically valid prospective cues* would be prospective but not predictive. Thus, anticipatory, 'feedforward', control can be prospective or predictive – motor commands based on predicted values (including internally estimated current values as opposed to feedback-delayed observed values) of the controlled variable would be predictive. As would motor commands based on predicted values of prospective cues, or indeed 'responding' to *any unobservable variables*.

In predictive control, the controller is genuinely a *model* of the controlled system, issuing information for feedforward control. See Lappi & Mole, 2018, for discussion in relation to the visual steering models literature.

And if this was not confusing enough, the terms ‘feedback’ and ‘feedforward’ can also be used in completely different way in the field of sensory neuroscience, where feedback means recurrent signals. I.e. feedback travels in the feedforward direction… See the end note → *Feedback and feedforward in the Brain*

---

[117] **PREDICTIVE CONTROL**

Predictive control as opposed to *prospective* control genuinely anticipates future states by some type of forward inference. That means it is *model-based control* operating on an internal model, a *representation* of the environment, rather than just *signals* generated by the environment. You could say that in a degenerate semantic sense the desired reference value in PID control represents a ‘goal’ state (making error comparison inverse inference) and ‘predicts’ the long-term behaviour of the plant (forward inference), but this way of looking at pure feedback control is somewhat procrustean, and pure feedback control is generally considered non-representational, *model-free control.*

Philosophically, genuine representation allows representation of unobserved variables, or even objectively non-existent features of the world (such as reward).

Model-based feedback/feedforward predictive control within control theory can give you an idea of how the brain may actually be using feedback – especially in tasks like race driving where it is clear that some kind of memory of the track environment is critically important. or simply to the predicted feedback – this is called ‘active inference’.

---

[118] Gross, 1998, p. 1445.

[119] This can also be happening to our driver… At the phenomenological level, when ‘in the zone’, time can seem to slow down. When *flow* is achieved, the fast, unforgiving pace of physical events is unreflected in the calm, measured state of conscious awareness. See end note on → *Flow*

[120] Most sports psychologists would probably identify the mental qualities of a racing driver *perceptual-cognitive expertise*, *decision making skill*, *mental toughness* or *resilience*, and by a host of other names.

Yet, there are intangibles that do not have a simple label – and of course merely giving something a *name* does not bring you any closer to understanding what that thing *is.* (See the parable of ‘the brown-throated thrush’ in Feynman (2005)).

---

[121] **SYNERGIES**

The term 'synergy' has several slightly different meanings in the literature, but generally what is meant are centrally activated molar patterns of fairly stereotypical behaviour involving multiple body segments and sensory organs. (Bernstein (1967, 1996). See also Ting & MacKay (2007), Profeta & Turvey (2018), Bruton & O'Dwyer, 2018).

The idea is that synergies simplify motor control because they can be selected (from a low–dimensional action representation), allowing complex context–sensitive adaptation to the details of a particular situation to be simplified. They have been proposed as combinatorial building blocks from which complex behavioural patterns are composed. At any given time, one or a few synergies could be recruited from a *restricted set of possible synergies*, which can be independently activated and weighted to adapt the motor pattern to the particular situation, creating *an efficient, sparse central encoding of motor efference.*

Synergies are invoked to explain how motor control may deal with the interrelated problems of so-called *curse of dimensionality*, arising from the *many degrees of freedom of the neuromuscular system* compared to the relatively *low dimensionality of the information needed to represent the motor goal* or trajectory, and ill–posedness of the inverse dynamic problem, i.e. finding a motor effector control signal that will generate a desired motor goal or trajectory (cf. the back-to-front logic of the racing line...).

In neuroanatomy and biomechanics (see Bizzi, Mussa–Ivaldi & Giszter, 1991; D'Avella, Saltiel & Bizzi, 2003; Ting & McKay, 2007), synergies refer to recurring stereotypical spatiotemporal patterns of motor activity when the co–activation of multiple muscle groups is:

1. *coordinated* and produces an *invariant motor outcome* (not in terms of trajectory, but in terms of a family of trajectories beginning from different initial state),
2. *balanced* in terms of agonist/antagonist activation, and
3. evoked in a *modular* manner by descending pathways innervating local populations of spinal or brain stem neurons.

Synergies have been most studied in the context of limb movement, but in visually guided tasks, like driving, the limbs and eyes must all be coordinated synergistically (see Lappi, 2016, for discussion).

---

[122] The use of 'waypoints' does not imply metronomic motor outputs at the

control level. This is a general property of human performance, not just related to racing. See discussion in **Chapter 5**.

[123] Lots of low-amplitude high-frequency modulations. It probably does not destabilize the car because the engine/drivetrain/tyres low-pass filter it – it does not mean he is not 'smooth': https://www.youtube.com/watch?v=JUVkVB3SUf4

It's not hesitation, either – and he does it in karts, too, and already before F1 so it's not turbo lag control habit.

This is all explained very well here by Driver61:

https://www.youtube.com/watch?v=N4kcLyYhThE

[124] Because the way our brain organizes actions into synergies is so deeply ingrained in how we move our body, I would wager that individual differences of this type would be quite difficult to modify.

Generally, the more deeply proceduralized skills become, the more repetition and feedback it takes to modify or unlearn them (and it's not at all clear that you should – what works for one individual may not be the best for another).

[125] On coordinate transformation computations in the brain related to visuomotor control, see Andersen et al. (1993), Pouget et al. (2002) and Crawford et al. (2011) and McGuire & Sabes (2009). For review and discussion of their importance for understanding locomotor tasks such as driving, see Lappi (2016).

[126] Optic flow is the motion of the visual field as the world seems to flow by you.

More technically it is the flow field of angular motions that visible texture patterns in a rigid environment project on a moving observer. See. Cutting (1986). Retinal flow is the registering of this pattern in a (moving) eye.

For discussion see Lappi (2014, 2016)

[127] In psychology and neuroscience this is known as *the binding problem*. Treisman (1996),

[128] *Dynamics* refers to the way forces set objects into motion, and how objects in motion exert forces on one another. *Mechanics* describes just the geometry of the motions, minus the forces. So describing the rotational and linear motions in suspension compression would be a mechanical description. Describing how a force, the vertical load on a tyre compresses the suspension would be a dynamical description.

[129] Even for the cortex the detailed connections are much more varied and not as strictly stratified as in the schematic diagram. Also, it is worth pointing out that the ostensibly 'lower' (old, 'mammalian brain') limbic cortex and many

subcortical structures reside above the evolutionarily newer parts of the cortex in this hierarchy.

See e.g. Fuster (2001, 2004), Chanes & Feldman Barrett (2016).

---

[130] FEEDBACK AND FEEDFORWARD IN THE BRAIN

Looked at from the point of view of feedback control, feedback signals flow 'bottom-up', recurrent signals in the sensory system, and efference copy of motor commands, flows 'top-down'. Feedforward motor commands in the motor system flow 'top-down' from motor areas to the motor periphery, and efference copy being sent from motor areas to sensory areas lower in the hierarchy.

Brain anatomists, however, tend to call sensory signals as being 'fed forward' and recurrent signals from higher in the hierarchy as being 'fed back'. So, the bottom-up sensory signals that for the engineer are feedback are 'feedforward signals' for the neuroscientist, and the recurrent signals and efference copy, the top-down signals. would be 'feedback signals' for the neuroscientist. Even if the efference copy is a copy of what a control engineer would call a feedforward signal! Confused? I know I was.

[131] These concepts can be, and have been, given a rigorous meaning in terms of Bayesian inference. See Friston (2009, 2010). The principles (i)-(iii) all build on predictive coding and fit with the general connectivity and experimental observations of brain function – although it must be said the bayesian brain hypothesis is still in the early stages of formation and cannot be considered definitively established fact.

---

[132] James (1890), p. 27: "A science of the relations of mind and brain must show how the elementary ingredients of the former correspond to the elementary functions of the latter."

---

[133] Okay maybe not the best example as so much seems nowadays to be concentrate in the chassis and engine number… which don't even do anything!

---

[134] **BRAIN IMAGING**

The insula is part of the interoceptive limbic system, and so-called salience network (Menon & Uddin, 2010). The modern analysis methods allow researchers to look at these interconnected 'intrinsic networks' of co-activating brain regions, rather than local 'foci' of activation.

One of the most important intrinsic networks for racing – also involved in the mental imagery and visualization task described next in the main text – is the frontoparietal attention network on the dorsal surface of the cortex. (Corbetta & Shulman, 2002; Ptak, 2012; Scolari et al., 2015)

[135] **Technical note** The activity difference can go in both directions, so 'activation' one can have also 'deactivation' (like one can have acceleration and deceleration, which are really just acceleration in opposite directions).

A highly technical note If the signal being looked at is the delivery of blood to the brain, and its extraction from the blood by neurons and support cells (as it is in functional fMRI), then there is a 'default' metabolic state that is quite consistent across a cerebral hemisphere when 'resting'.

When a brain area becomes 'activated' in fMRI terms, regional blood flow increases bringing in oxygenated blood and more oxygen that the tissue uses, leading to higher rate of hemoglobin with oxygen to hemoglobin without oxygen; this blood oxygenation level is what the fMRI signal picks up. The 'resting' state can be considered a goldilocks state where the blood delivers just the right amount of oxygen to all cerebral tissues, yielding a consistent ratio of oxygenated to non-oxygenated blood. (The ratio is called by the fancy name oxygen extraction fraction – the amount of blood and oxygen used by different regions can vary wildly, though). In this case the baseline reference level can be set to be the hemispheric average oxygen extraction rate, and activation of *all* cortical areas can be compared to this *same* standard, rather than doing an area-wise comparison as in the example we run in the main text.

The general point that any local activation in any task A results from a statistical comparison of brain function during task A to some control task (the control task here being 'resting').

Brain areas that consistently *deactivate* relative to this baseline when a person is engaged in any of a variety of attentionally demanding tasks are jointly called 'the brain's default mode network', by the way (Raichle et al. 2001; Raichle, 2015).

What happens during 'the default mode'? (What does the brain do, in cognitive terms, when you do not tell the person lying in the scanner to do or think about anything in particular?). This is a matter of continuing debate since the discovery of the default mode network twenty years ago. This is an example of the problem when the task A and task B being compared are not very neatly defined ('resting' vs. 'doing everything else'). Here there is perhaps a difference between the cognitive approach to brain function, from which point of view the 'resting' state is an unsatisfactory control state (because of potentially high intra- and interindividual variability), and a neurophysiological approach by which it is quite satisfactory (metabolically consistent within and between individuals).

A *super* technical note, *are you still reading this*?! A network defined in terms of 'intrinsic activity' this way – by reference to a global baseline – is not necessarily the same thing as an 'intrinsic network' more generally defined by functional connectivity (e.g. interarea synchronization of brain activity) during 'rest' or a 'cognitive task' (Bullmore & Sporns, 2009; Yeo et al., 2011). How well networks defined and identified in different way match and overlap in an empirical question.

All of this makes absolutely no difference to any of the arguments elsewhere in this book, but is mentioned here to match the anorak level of some of the notes on racecar engineering and driving technique, and presented for the benefit or otherwise of all brain science students out there reading this...

[136] Large-scale functional networks can be defined on a number of criteria going beyond direct anatomical connections, such as resting state metabolism, synchronization of electric activity, or joint engagement in a specific task.

[137] Brain imaging research on driving are reviewed in Lappi (2015) and Navarro et al. (2019). There is also one brain imaging study of racing drivers (Bernardi et al., 2014), also reviewed and put into the context of the Mcruer hierarchy in Lappi, (2015).

---

[138] META-ANALYSIS OF BRAIN ACTIVATION RELATED TO DRIVING

Meta-analyses use statistical methods to pool together data from multiple studies to assess which activation loci are seen consistently across a larger sample of data than can be acquired in one study. One challenge in meta-analyses is that the choice of which experiments' data to pool has to take into account what *exact* task the participants were performing, and what control task data was used to get the reported 'activation' contrasts. (See the next end note).

The authors of this particular meta analysis classified tasks used in the original brain imaging studies into *strategic*, *tactical*, and *operational*, and performed separate meta analyses for each subclass of experiments to identify foci of activation consistently seen so far in each type of experiment. (This classification, commonly used in traffic psychology, partly aligns with the navigation/guidance/control classification we have used here, although it is more general but less well defined).

They then compared the activation likelihood maps across categories to identify possible substrates of decisions cascading down from a higher level to a lower one:

'*it is proposed that at any time while driving the strategical level is engaged at first, then if required, the situation is processed at the tactical level. Finally, the decisions made at the tactical level are translated into motor programs at the operational level when required*' (p. 475).

According to their hypothesis, the highest level, more explicit conscious processes associated with the anterior cerebrum would monitor the situation and cascade control to more implicit lower levels (posterior cerebrum), if action is deemed necessary, which then perform automatic habitual actions.

[139] A problem with interpreting brain imaging literature is that the specific task identified as 'driving' is not really the same across studies, and also the choice of what 'non-driving' task the statistical contrast is made against is not always the same.

This reflects in part the immaturity of the field, in lacking a clear task analysis that would allow a more fine-grained comparison of results. It is not specific to studying driving, however, but rather endemic to the brain imaging tradition, where the characterization of the experimental and control tasks is often not a detailed, ideally computational, analysis of the operations each task requires, allowing one to pinpoint exactly which operations are different in the two tasks. As one major textbook in cognitive neuroscience puts it: '*a model of the neural instantiation of a cognitive function depends critically on the validity of the function it seeks to explain*' (Postle, 2015)

There is a chicken-and-egg problem here, of course, in that the brain activation results can only be interpreted if one already has in hand a detailed theory of the mental functions. On the other hand, many wish we could bootstrap cognitive analyses to the brain research, in order to give a more solid grounding to hypothetical cognitive task analyses…

[140] These are the categories of Michon (1985) – *operational-tactical-strategic* – which is widely used in traffic psychology. This classification is broadly similar to the McRuer (1977) hierarchy of *control-guidance-navigation* used here, though the levels do not align exactly.

The traffic psychological classification of 'driving behaviours' is also quite broad, and because it is based on only intuitive assessment and verbal description, more fuzzy in definition. In terms of intuitively classifying brain loci there is probably not much to choose between them, though.

Levels of the Michon hierarchy,

as used by Navarro et al. (2019)

Strategic	Driving goals
	Route planning
	Navigation
Tactical	Overtaking
	Lane selection
	Exiting highway
Operational	Steering
	Accelerating
	Braking

The Michon hierarchy is perhaps more suited to traffic psychology, however, because in it you can more readily classify also traffic behaviours other than strictly *driving* – e.g. waving at another road user in an intersection may be tactical, but not really guidance or navigation. In terms of a detailed task analysis for driving, though, the McRuer framework is better suited because it is more precise and unambiguous to relate to mechanistic hypotheses in engineering and neuroscience (See discussion in Lappi, 2015 and Lappi & Mole, 2018).

Because 'driving in traffic' is not a well-defined task, comprising of many tasks that go beyond our interest in 'pure driving' it may be instructive also to consider brain activation restricted basic steering tasks (i.e. with no complex secondary tasks):

The most common finding is activation of premotor frontal areas and occipital visual areas extending into parietal areas (cuneus, precuneus), all the way to posterior parietal cortex. The frontal activation extends from premotor to prefrontal cortex and the insula. Also activation in posterior cingulate/ retrosplenial cortex, in the right hemisphere in particular, is commonly reported. Cerebellar activation at various sites has also been observed in most studies. (Lappi, 2015; Lappi & Mole, 2018).

---

[141] Bernardi et al., 2014. For further theoretical analysis of the results, and how they relate to the brain imaging literature on normal driving, see Lappi, 2015).

[142] ISC does not measure activation relative to a predetermined stimulus event, but compares the signal time series of individuals at a given locus.

Loci where the time series are correlated, i.e. the signal changes in step in different subjects, are said to be activated, so activation indicates synchrony of brain activity between individuals. See Hasson et al., (2010); Nummenmaa et al., (2018); Nastase et al. (2019). The brain 'activation' is not calculated as a 'response' to specific stimulus events.

[143] Before-after training comparisons in brain imaging studies of neuroplasticity have shown that learning a motor skill can change the size of the motor cortex. In response to learning to play an instrument the relevant bits of the brain increase in size in musicians' brains. For review see Münte et al., (2002).

[144] There are aspects that are maybe more hardwired, such as the organization of pace, trot and gallop in spinal cord reflexes/synergies (see end note on → *Synergies*).

[145] This is obviously a sci-fi reference to Edgar Rice Burroughs' John Carter books... just wanted to exhibit some level of nerdiness here...

---

[146] ACTIVE PERCEPTION

Active perception refers to how we actively move our sensory organs, and how this affects incoming sensory signals (feedback) to the benefit of perceptual analysis. This concept should not be confused with *active sensing* in the engineering sense, where sensors such as sonar and radar are 'active' in that they send out signals and energy to the environment and then sense the returning echo.

Palpation, where we use our extended body schema to sense the texture of a surface, is also an example of active perception, and the hand is an active sensor. But sampling visual scenes by eye movements is also active perception.

The eye is used in active perception because the brain controls the movements of the eye in order to eye to see; but it is not an active sensor because it only registers ambient patterns of light – it does not send out any light of its own.

On active perception see Ballard (1991) and Fermuller & Aloimonos (1995). For recent neuroscience applications to modelling vision and somatosensation see e.g. Raudies et al. (2012), Kuang et al. (2012), Ahissar & Arieli (2012).

---

[147] Of course, there are differences to playing a musical instrument and operating a vehicle as well.

First, in terms of how the performance is organized in time. In music the tempo is set, and the musician must not try to go any faster – the driver is free to set their speed (sort of) and may decide to go faster or slower at any point in time. How fast things should be done is limited by vehicle response and driver reaction latencies – this is typically not the case in music.

Second, driving unfolds in space as well as in time at many levels of scale: in addition to how you move your hands and feet to operate the instrument, there is the way the instrument itself moves through large-scale space.

Finally, mistakes are embarrassing and irritating in both cases, but in race driving there is an added element of physical harm; so taking liberties and trying 'something completely different' is much more severely restricted.

[148] This is based on https://www.youtube.com/watch?v=BBz-Jyr23M4

[149] One-shot learning is actually a technical term in machine learning. You can even go one better with zero-shot learning!

[150] The rate of information in the automotive context is discussed and analysed in Senders et al. (1967) and Kujala et al. (2015).

[151] This does seem set to continue; perhaps inadvertently encouraged by 'driver assistance' and autonomous technologies that lets people get away with it (effectively removing sources of negative feedback, which would be needed to make people discontinue such behaviours).

[152] Keele (1968); Schmidt (1975); McRuer et al. (1977). See Schmidt (2003), Shea & Wulf (2005) and Summers & Anson (2009) for more contemporary reviews of the precognitive motor program idea, and Lappi & Mole (2018) for discussion pertaining to driver modelling.

[153] Trying to increase one's powers of anticipation by longer, more elaborate *ballistic* motor programs would be almost self-defeating: the longer the motor program takes to run, the more head start the brain is trying to give itself, the bigger the variability of timing at the end of the program would likely be.

This is because in neurological processes noise tends to scale with duration, and other magnitudes. This is called scalar variability, or Weber's law.

For a predictive brain (generative, bayesian) view of scalar variability and other classical psychophysiological results, see Petzschner et al. (2015).

[154] Miall & Wolpert, 1996; Wolpert & Kawato, 1998; for a historical overview see Desmurget & Grafton, 2000.

[155] Mumford, 1992; Rao & Ballard, 1999; Friston, 2009; 2010; Clark, 2013.

All these different aspects of the mind fall under a single computational principle: prediction error minimization. At the level of concrete mechanisms and tasks, there is likewise no sharp piecemeal division into perception, attention, learning and memory (cognition), motor control, etc. This makes it possible to develop models that could account, in quantitative detail, for complete task performance in representative natural tasks.

[156] See https://quoteinvestigator.com/2018/04/14/theory/#note-18386-1 for the origin of this famous quote; it has been attributed to Albert Einstein, Richard Feynman and Yogi Berra, among others. The original seems to be from Brewster (1882), p. 202 – although in slightly different, less condense form.

[157] Newell (1973).

---

[158] THE ECOLOGICAL APPROACH TO COGNITIVE SCIENCE

I believe that understanding human behaviour, brain function and the mind can only be achieved by taking a more naturalistic approach than is common in behavioral sciences nowadays. There is a tendency in the academia to consider study of real-world behaviour –things people actually *do* – as 'applied'. But if one views an ecological approach to human cognition as just task-specific 'application' of supposedly 'general' facts discovered in the lab, I fear one's look at the puzzle of human behaviour and brain function is stunted.

Experimental control is important; but it is also widely recognized that often experimental tasks of questionable ecological validity. The problem is not when experiments are not done 'in the wild', it's when they are not representative of the challenges and skills they are intended to capture (e.g. because of a need to make them simple, 'modelling friendly', easier to analyse). There is then a danger to side-step the problem of actually trying to *understand* the real behaviour you were originally supposed to be able to generalize to.

The alternative is to go through the trouble of trying to understand the ecology of real behaviour from multiple points of view (like the cognition and brain function and physics of driving!), and actually to try to derive experimental paradigms from careful (quantitative) observations of real behaviour.

As Gibson (1986, p.3) put it:

"It is not true that 'the laboratory can never be like life'.

The laboratory must be like life!"

---

[159] Taruffi (1959), p.115.

REFERENCES

Ahissar, E. & Arieli, A. (2012). Seeing via Miniature Eye Movements: a Dynamic Hypothesis for Vision. *Frontiers in Computational Neuroscience*, 6, 89, 1-27. doi: 10.3389/fncom.2012.00089

Ahissar, E. & Assa, E. (2015). Perception as a Closed-loop Convergence Process. *eLife*, 5:e12830. doi: 10.7554/eLife.12830

AlphaStar Team, 2019, AlphaStar: Mastering the Real-Time Strategy Game StarCraft II, blog post, retrieved from https://deepmind.com/blog/article/alphastar-mastering-real-time-strategy-game-starcraft-ii

Andersen, R. A., Snyder, L. H., Li, C. S., & Stricanne, B. (1993). Coordinate Transformations in the Representation of Spatial Information. *Current Opinion in Neurobiology*, 3(2), 171-176.

Backman, J., Häkkinen, K., Ylinen, J., Häkkinen, A., & Kyröläinen, H. (2005). Neuromuscular Performance Characteristics of Open-wheel and Rally drivers. *Journal of Strength and Conditioning Research*, 19(4), 777-784.

Badre, D. (2008). Cognitive Control, Hierarchy, and the Rostro-caudal Organization of the Frontal Lobes. *Trends in Cognitive Sciences*, 12(5), 193-200. doi:10.1016/j.tics.2008.02.004

Ballard, D.H. (1991). Animate Vision. *Artificial Intelligence*, 48, 57–86.

Baur, H., Müller, S., Hirschmüller, A., Huber, G., & Mayer, F. (2006). Reactivity, Stability, and Strength Performance Capacity in Motor Sports. *British Journal of Sports Medicine*, 40(11), 906-911.

Beikmann, R. (2015). *Physics for Gearheads. An Introduction to Vehicle Dynamics, Energy and Power with Examples from Motorsports*. Bentley Publishers.

Bentley, R. (1998). *Speed Secrets: Professional Race Driving Techniques*. Motorbooks, an imprint of MBI Publishing Company.

Bentley, R. (2003). *Speed Secrets 2: More Professional Race Driving Techniques*. Motorbooks, an imprint of MBI Publishing Company

Bereiter, C., & Scardamalia, M. (1993). *Surpassing ourselves. An Inquiry into the Nature and Implications of Expertise*. Open Court.

Bernardi, G., Ricciardi, E., Sani, L., Gaglianese, A., Papasogli, A., Ceccarelli, R., ... & Pietrini, P. (2013). How Skill Expertise Shapes the Brain Functional Architecture: an fMRI Study of Visuo-spatial and Motor Processing in Professional Racing-car and Naïve Drivers. *PloS ONE*, 8(10), e77764

Bernardi, G., Cecchetti, L., Handjaras, G., Sani, L., Gaglianese, A., Ceccarelli, R., ... & Pietrini, P. (2014). It's Not All in yYour Car: Functional and Structural Correlates of Exceptional Driving Skills in Professional Racers. *Frontiers in human neuroscience*, *8*, 888, 1-12.

Bernardin, D., Kadone, H., Bennequin, D., Sugar, T., Zaoui, M., & Berthoz, A. (2012). Gaze Anticipation During Human Locomotion. *Experimental Brain Research*, 223(1), 65-78.

Bernstein, N.A., 1967. *The Co-ordination and Regulation of Movements*. PergamonPress, New York.

Bernstein, N. A. (1996). On the Construction of Movements. In M. L. Latash, & M. Turvey (Eds.). *Dexterity and its Development* (pp. 3–244). Mahwah, NJ: Lawrence Erlbaum.

Bizzi, E., Mussa-Ivaldi, F.A., Giszter, S., 1991. Computations Underlying the Execution of Movement: a Biological Perspective. *Science* 253 (5017), 287–291

Botvinick, M. M. (2008). Hierarchical Models of Behaviour and Prefrontal Function. *Trends in Cognitive Sciences*, 12(5), 201–208.

Bullmore, E., & Sporns, O. (2009). Complex Brain Networks: Graph Theoretical Analysis of Structural and Functional Systems. *Nature Reviews Neuroscience*, *10*(3), 186-198.

Brewster, B. (1882). Portfolio: Theory and Practice *The Yale Literary Magazine, Conducted by the Students of Yale College*, Volume 47, Number 5, New Haven, Connecticut.

Brown, J.W.H, Stanton, N.A. & Revell, K.A. (2020). The Evolution of Steering Wheel Design in Motorsport. In: P.M. Salmon, S. McLean, C. Dallat, ... & A. Hulme (eds.): *Human Factors and Ergonomics in Sport*, CRC Press.

Bruton, M., & O'Dwyer, N. (2018). Synergies in Coordination: a Comprehensive Overview of Neural, Computational, and Behavioural Approaches. *Journal of Neurophysiology*, *120*(6), 2761-2774.

Buckner, R.L. (2013). The Cerebellum and Cognitive Function: 25 Years of Insight from Anatomy and Neuroimaging. *Neuron*, 30, 807-815. doi: 10.1016/j.neuron.2013.10.044

Buhusi, C.V. & Meck, W.H. (2005). What Makes Us Tick? Functional and Neural Mechanisms of interval Timing. *Nature Reviews Neuroscience*, 6, 755 -765. doi: 10.1038/nrn1764

Bullmore, E. & Sporns, O. (2009). Complex Brain Networks: Graph Theoretical Analysis of Structural and Functional Systems. *Nature Reviews Neuroscience*, 10, 186-198.

Burke, R.E. (2007). Sir Charles Sherrington's The Integrative Action of the Nervous System: A Centenary Appreciation. *Brain*, 130, 887-894. doi:10.1093/brain/awm022

Burr, D.C., Morrone, M.C. (2012). Constructing Stable Spatial Maps of the World. *Perception*, 41, 1355 – 1372. doi:10.1068/p7392

Buzsaki, G., Geisler, C., Henze, D.A. & Wang, X.J. (2004). Circuit Complexity and Axon Wiring, Economy of Cortical Interneurons. *Trends in Neurosciences*, 27, 4, 186-193. doi:10.1016/j.tins.2004.02.007

Chanes, L., & Feldman Barrett, L. (2016). Redefining the Role of Limbic Areas in Cortical Processing. *Trends in cognitive sciences*, *20*(2), 96-106.

Chase, W.G. & Simon, H.A. (1973). Perception in Chess. *Cognitive Psychology*, 4, 55-81.

Chater, N. (2018). *The Mind is Flat. The Illusion of Mental Depth and the Improvised Mind*. Pengiuin Random house.

Cherniak, C. (1994). Component Placement Optimization in the Brain. *The Journal of Neuroscience*, , 14(4): 2418-2427.

Clark, A. 2013. 'Whatever Next? Predictive Brains, Situated Agents, and the Future of Cognitive Science.' *Behavioural and Brain Sciences 36*(3): 181–204. doi:10.1017/S0140525X12000477.

Clark, A. (2015). *Surfing uncertainty: Prediction, Action, and the Embodied Mind.* Oxford University Press.

Clayton, M. S., Yeung, N., & Kadosh, R. C. (2015). The Roles of Cortical Oscillations in Sustained Attention. *Trends in cognitive sciences*, *19*(4), 188-195.

Code, K. (1983). *A Twist of the Wrist – The Motorcycle Road Racer's Handbook*. Code Break.

Code, K. (1993). *A Twist of the Wrist Volume II*. Code Break.

Conway, A. R., Kane, M. J., & Engle, R. W. (2003). Working Memory Capacity and Its Relation to General Intelligence. *Trends in Cognitive Sciences, 7*(12), 547-552. doi:10.1016/j.tics.2003.10.005

Cooper, R. & Shallice, T. (2000). Contention Scheduling And the Control of Routine Activities. *Cognitive Neuropsychology,* 17(4), 297–338.

Corbetta M, Shulman GL. (2002). Control of Goal-directed and Stimulus-driven Attention in the Brain. *Nature Reviews Neuroscience*. 3(3):201-15.

Cutting, J. E. (1986). *Perception with an Eye for Motion* (Vol. 1). Cambridge, MA: MIT Press.

Crawford, J. D., Henriques, D. Y., & Medendorp, W. P. (2011). Three-dimensional Transformations for Goal-directed Action. *Annual Reviews of Neuroscience*, 34, 309-331.

Crundall, E., Crundall, D., & Stedmon, A. W. (2012). Negotiating Left-hand and Right-hand bends: A Motorcycle Simulator Study to Investigate Experiential and Behaviour Differences across Rider Groups. *PLoS One*, *7*(1), e29978.

Csikszentmlhalyi, M. (1975/2000). *Beyond Boredom and Anxiety*. Jossey-Bass.

Csikszentmihalyi, M. (2014). *Flow and the Foundations of Positive Psychology: The Collected Works of Mihaly Csikszentmihalyi*. Springer. doi: 10.1007/978-94-017-9088-8_17

d'Avella, A., Saltiel, P., & Bizzi, E. (2003). Combinations of Muscle Synergies in the Construction of a Natural Motor Behaviour. *Nature Neuroscience*, *6*(3), 300-308.

Dayan, P., & Niv, Y. (2008). Reinforcement learning: the good, the bad and the ugly. *Current opinion in neurobiology*, 18(2), 185-196.

Daye, P. M., Optican, L. M., Blohm, G., & Lefevre, P. (2014). Hierarchical Control of Two-dimensional Gaze Saccades. *Journal of Computational Neuroscience*, 36(3), 355-382.

De Groot, A. D. (1946/1965). *Thought and Choice in Chess.* Haag: Mouton Publishers.

De Groot, A. D. & Gobet, F. (1996). *Perception and Memory in Chess.* Assen: Van Gorcum.

Dennett, D. C. (2005). Cognitive wheels: The frame problem of AI. *Language and Thought*, *3*, 217.

Desmurget, M., & Grafton, S. (2000). Forward Modeling Allows Feedback Control for Fast Reaching Movements. *Trends in Cognitive Sciences*, 4(11), 423-431.

Diamond, A. (2013). Executive Functions. *Annual Review of Psychology, 64*, 135-168. doi: 10.1146/annurev-psych-113011-143750

Dickinson, M.H., Farley, C.T., Full, R.J., Koehl, M.A., Kram, R. & Lehman, S. (2000). How Animals Move: An Integrative View. *Science,* 288: 100–106. doi: 10.1126/science.288.5463.100

Didierjean, A. & Gobet, F. (2008). Sherlock Holmes – an Expert's View of Expertise. *British Journal of Psychology*, 99, 109–125. DOI:10.1348/000712607X224469

Donges, E. (1978). A Two-level Model of Driver Steering Behaviour. *Human Factors*, 20, 691–707. doi: 10.1177/001872087802000607

Donohue, M. & vanValkenburgh, P. (1975). *The Unfair Advantage*. Bentley Publishers.

Dove, T. (2016). *Learn How to Master the Art of Kart Driving*. EvenFlow Kart Driver Coaching.

Dreyfus, H.L. (1992). *What Computers Still Can't Do*, MIT Press.

Elford, V. (2008). *Porsche High-Performance Driving Handbook, 2nd edition*. Motorbooks, an imprint of MBI Publishing Company.

Endsley M.R. (1995). Toward a Theory of Situation Awareness in Dynamic Systems. *Human Factors*, 37(1), 32-64.

Engle, R. W. (2002). Working Memory Capacity as Executive Attention. *Current Directions in Psychological Science, 11*(1), 19-23.

Ericsson, K.A. (2003). Protocol Analysis and Expert Thought: Concurrent Verbalizations of Thinking during Experts' Performance on Representative Tasks. In: Ericsson, K. A., Charness, N., Feltovich, P. J., & Hoffman, R. R. (Eds.). (2006). *The Cambridge Handbook of Expertise and Expert Performance.* Cambridge University Press, pp. 685-705.

Ericsson, K. A., Hoffman, R. R., Kozbelt, A., and Williams, A. M. (eds.) (2018). *The Cambridge Handbook of Expertise and Expert* Performance. Cambridge University Press.

Ericsson, K. A., Krampe, R. T., and Tesch-Römer, C. (1993). The role of deliberate practice in the acquisition of expert performance. *Psychological Review*, 100, 363–406. doi: 10.1037/0033-295X.100.3.363

Ericsson, K.A. (2020). Given That the Detailed Original Criteria for Deliberate Practice Have Not Changed, Could the Understanding of this Complex Concept Have Improved Over Time? A Response to Macnamara and Hambrick (2020). *Psychological Research.* doi: 10.1007/s00426-020-01368-3

Ericsson, K.A. & Pool, R. (2017). *Peak. Secrets from the New Science of Expertise.* Houghton Mifflin Harcourt, Boston.

Feldman, J. & Ballard, D. (1982). Connectionist Models and their Properties. *Cognitive Science*, 6, 205-254.

Fermüller, C. & Aloimonos, Y. (1995). Vision and Action. *Image and Vision Computing*, 13(10), 725–744.

Friston, K. (2009). The Free-energy Principle: a Rough Guide to the Brain?. *Trends in Cognitive Aciences*, *13*(7), 293-301.

Friston, K. (2010). The Free-energy Principle: a Unified Brain Theory?. *Nature Reviews Neuroscience*, *11*(2), 127-138.

Gazzaniga, M. S. (2013). Shifting Gears: Seeking New Approaches for Mind/brain Mechanisms. *Annual Review of Psychology*, 64, 1-20.

Gladwell, M. (2008). *Outliers, the Story of Success*. Little, Brown.

Glaser, R. (1985). *Thoughts on Expertise.* Technical Report No. 8. University of Pittsburgh Learning and Development Center.

Gleick, J. (1987). *Chaos, Making of a New Science.*

Fermüller, C. & Aloimonos, Y. (1995). Vision and Action. *Image and Vision Computing*, 13(10), 725.744.

Feynman, R. P. (2005). *The Pleasure of Finding Things Out: The Best Short Works of Richard P. Feynman.* Basic Books.

Freestone, D.M. & Church, R.M. (2016). Optimal Timing. *Current Opinion in Behavioural Sciences*, 8, 276–281. doi: 10.1016/j.cobeha.2016.02.031

Friston, K., Mattout, J. & Kilner, J. (2011) Action Understanding and Active Inference. *Biological Cybernetics 104*:137–60.

Friston, K., Rigoli, F., Ognibene, D., Mathys, C., Fitzgerald, T., & Pezzulo, G. (2015). Active inference and Epistemic Value. *Cognitive neuroscience, 6*(4), 187-214.

Fuchs, F., Song, Y., Kaufmann, E., Scaramuzza, D., & Duerr, P. (2020). Super-Human Performance in Gran Turismo Sport Using Deep Reinforcement Learning. *arXiv preprint arXiv:2008.07971.*

Fuster, J.M. (2001). The Prefrontal Cortex—An Update: Time Is of the Essence. *Neuron*, 30, 319–333.

Fuster, J.M. (2004). Upper Processing Stages of the Perception–action Cycle. *Trends in Cognitive Sciences*, 8, 4, 143-145. doi:10.1016/j.tics.2004.02.004

Gibson, J. J. (1986). *The Ecological Approach to Visual Perception*. Hillsdale, NJ: Lawrence Erlbaum Associates.

Gilchrist, A. L. (2015). How Should We Measure Chunks? A Continuing Issue in Chunking Research and a Way Forward. *Frontiers in psychology, 6*, 1456.

Gobet, F. & Simon, H. A. (1996a). Recall of Random and Distorted Positions: Implications for the Theory of Expertise. *Memory & Cognition*, 24, 493–503.

Gobet, F. & Simon, H. A. (1996b). Recall of Rapidly Presented Random Chess Positions is a Function of Skill. *Psychonomic bulletin & review*, 3, 159–163.

Gobet, F. & Simon, H. A. (1996c). Templates in chess memory: A mechanism for recalling several boards. *Cognitive psychology*, 31, 1–40.

Gobet, F. (1998). Expert Memory: A Comparison of Four Theories. *Cognition*, 66, 115–152.

Gobet, F., Lane, P. C., Croker, S., Cheng, P. C., Jones, G., Oliver, I., & Pine, J. M. (2001). Chunking Mechanisms in Human Learning. *Trends in cognitive sciences*, *5*(6), 236-243.

Gobet, F., Lloyd-Kelly, M., & Lane, P. C. (2016). What's in a Name? The Multiple Meanings of 'Chunk' and 'Chunking'. *Frontiers in psychology, 7*, 102.

Gobet, F. & Simon, H. A. (1998). Expert Chess Memory: Revisiting the Chunking Hypothesis. *Memory*, 6, 225–255.

Grasso, R., Glasauer, S., Takei, Y., & Berthoz, A. (1996). The Predictive Brain: Anticipatory Control of Head Direction for the Steering of Locomotion. *Neuroreport*, 7(6), 1170-1174.

Grasso, R., Prévost, P., Ivanenko, Y. P., & Berthoz, A. (1998). Eye-head Coordination for the Steering of Locomotion in Humans: an Anticipatory Synergy. *Neuroscience Letters*, 253(2), 115-118.

Gross, C. G. (1987). Early history of Neuroscience. *Encyclopedia of Neuroscience, 2,* 843-846.

Harrington, D.L. & Jahanshahi, M. (20i6). Reconfiguration of Striatal Connectivity for Timing and Action. *Current Opinion in Behavioural Sciences*, 8, 78–84. doi: 10.1016/j.cobeha.2016.02.007

Harris, D., Allen, K., Vine, S., Wilson, M., 2020. A Systematic Review and Meta-analysis of the Relationship between Flow States and Performance. *PsyArXiV*. doi:10.31234/osf.io/qg852.

Hasson, U., Malach, R., & Heeger, D. J. (2010). Reliability of Cortical Activity during Natural Stimulation. *Trends in cognitive sciences*, *14*(1), 40-48.

Heathcote, A., Brown, S., & Mewhort, D. J. (2000). The Power Law Repealed: The case for an Exponential Law of Practice. *Psychonomic bulletin & review*, *7*(2), 185-207.

Hebb, D. O. (1955). Drives and the CNS (Conceptual Nervous System). *Psychological Review*, *62*(4), 243-254.

Hilton, C. (2010). *The Toleman Story: Last Romantics in Formula 1*. Veloce.

Hirst, W., & Phelps, E. A. (2016). Flashbulb Memories. *Current Directions in Psychological Science*, *25*(1), 36-41.

Holbert, A., Holbert, B. & Bochroch, A. (1982). *Driving to Win!* Aztex Corporation.

Hunt, A. R., & Kingstone, A. (2003). Covert and Overt Voluntary Attention: Linked or Independent?. *Cognitive Brain Research*, *18*(1), 102-105. doi: 10.1016/j.cogbrainres.2003.08.006

Imai, T., Moore, S. T., Raphan, T., & Cohen, B. (2001). Interaction of the Body, Head, and Eyes During Walking and Turning. *Experimental Brain Research*, 136(1), 1-18.

Jackson, S. A. (1996). Toward a Conceptual Understanding of the Flow Experience in Elite Athletes. *Research quarterly for exercise and sport*, *67*(1), 76-90.

Jagacinski, R.J. & Flach, J.M.(2002). *Control Theory for Humans: Quantitative Approaches To Modeling Performance*. Lawrence Erlbaum Associates Inc.

James, W. (1890). *The Principles of Psychology*. Macmillan.

Jenkinson, D.S. (1959/1997) *The Racing Driver – the Theory and Practice of Fast Driving.* Bentley Publishers.

Jeon, H. (2014). Hierarchical Processing in the Prefrontal Cortex in a Variety of Cognitive Domains. *Frontiers in Systems Neuroscience*, 8, 223, 1-8. doi: 10.3389/fnsys.2014.00223

Keele, S. W. (1968). Movement Control in Skilled Motor Performance.*Psychological Bulletin*, *70*(6 Part 1), 387-403.

Keil, F.C. (2003) Folkscience: Coarse iInterpretations of a Complex Reality. *Trends in Cognitive Sciences*, 7, 368-373. doi:10.1016/S1364-6613(03)00158-X

Koechlin, E. & Summerfield, C. (2007). An Information Theoretical Approach to Prefrontal Executive Function. *Trends in Cognitive Sciences*, 11, 6,229-235. doi:10.1016/j.tics.2007.04.005

Kuang X, Poletti M, Victor JD, Rucci M (2012) Temporal Encoding of Spatial Information During Active Visual Fixation. Current Biology, 22(6): 510- 514. doi: 10.1016/j.cub.2012.01.050

Kujala, T., Mäkelä, J., Kotilainen, I. & Tokkonen, T. (2015). The Attentional Demand of Automobile Driving Revisited. *Human Factors*, 58 (1), 163 – 180. doi: 10.1177/0018720815595901

Kücükdurmaz, F. (2012). Driver as a High Level Athlete. In: M.N. Doral et al. (eds.), *Sports Injuries*. Berlin Heidelberg: Springer-Verlag. DOI: 10.1007/978-3-642-15630-4_149

Land, M. F. (1992). Predictable Eye-head Coordination during Driving. *Nature*, 359, 318-320.

Land, M. F. (2006). Eye Movements and the Control of Actions in Everyday Life. *Progress in Retinal and Eye Research*, 25(3), 296-324. doi:10.1016/j.preteyeres.2006.01.002

Land, M.F. & Furneaux, S. (1997). The Knowledge Base of the Oculomotor System. *Philosophical Transactions of the Royal Society B*, 352: 1231–1239.

Land, M. F., & Tatler, B. W. (2001). Steering with the Head: The Visual Strategy of a Racing Driver. *Current Biology*, 11(15), 1215-1220.

Lappi, O., & Lehtonen, E. (2013). Eye-movements in Real Curve driving: Pursuit-like Optokinesis in Vehicle Frame of Reference, Stability in an Allocentric-reference Coordinate System. *Journal of Eye Movement Research*, 6(1):4, 1–13.

Lappi, O., Pekkanen, J., & Itkonen, T. H. (2013). Pursuit Eye-movements in Curve driving Differentiate between Future Path and Tangent Point Models. *PLOS ONE*, 8(7), e68326.

Lappi, O. (2013). *Eyes on the Road: Eye Movements and the Visual Control of Locomotion in Curve Driving*. (Studies in Cognitive Science 5). University of Helsinki.

Lappi, O. (2014). Future Path and Tangent Point Models in the Visual Control of Locomotion in Curve Driving. *Journal of Vision*, 14, 21. doi: 10.1167/14.12.21

Lappi, O. (2015). The Racer's Brain: How Domain Expertise is Reflected in the Neural Substrates of Driving. *Frontiers in Human Neuroscience*, 9:635. doi: 10.3389/fnhum.2015.00635

Lappi, O. (2016). Eye Movements in the Wild: Oculomotor Control, Gaze Behaviour & Frames of Reference. *Neuroscience & Biobehavioural Reviews, 69,* 49-68.

Lappi, O. (2018). The Racer's Mind—How Core Perceptual-Cognitive Expertise Is Reflected in Deliberate Practice Procedures in Professional Motorsport. *Frontiers in Psychology*, 12, 1294. 1-20. doi: 10.3389/fpsyg.2018.01294

Lappi, O. (2022). Gaze Strategies in Driving–An Ecological Approach. *Frontiers in Psychology,* 13, 821440. doi: 10.3389/fpsyg.2022.821440

Lappi, O. (2022). Egocentric Chunking in the Predictive Brain: A Cognitive Basis of Expert Performance in High-Speed Sports. *Frontiers in Human Neuroscience*, 12, 8226887. doi: 10.3389/fnhum.2022.822887

Lappi, O., Rinkkala, P., Pekkanen, J. (2017). Systematic Observation of an Expert Driver's Gaze Strategy – An On-road Case Study. *Frontiers in Psychology*, 8, 620, 1-13. doi: 10.3389/fpsyg.2017.00620

Lappi. O. & Mole, C.D. (2018). Visuomotor Control, Eye Movements, and Steering: A Unified Approach for Incorporating Feedback, Feedforward, and Internal Models. *Psychological Bulletin*. 144(10), 981-1001. doi:10.1037/bul0000150

Lappi, O., Pekkanen, J., Rinkkala, P., Tuhkanen, S., Tuononen, A., & Virtanen, J. P. (2020). Humans Use Optokinetic Eye Movements to Track Waypoints for Steering. *Scientific reports*, *10*(1), 1-14.

Lashley, K. S. (1951). The Problem of Serial Order in Behaviour. In L. A. Jeffress (Ed.) *Cerebral Mechanisms and Behaviour.* John Wiley and sons, New York, pp. 112–136.

Lee, T. S. & Mumford, D. (2003). Hierarchical Bayesian inference in the visual cortex. *Journal of Optical Society of America*, A 20(7):1434–48.

Lima, I. R., Haar, S., Di Grassi, L., & Faisal, A. A. (2020). Neurobehavioural Signatures in Race Car Driving: a Case Study. *Scientific reports*, *10*(1), 1-9.

Logan, G. D. (1988). Toward an Instance Theory of Automatization. *Psychological Review*, 95, 492–527. doi: 10.1037/0033-295X.95.4.492

Loomis, J. M., Klatzky, R. L., & Giudice, N. A. (2013). Representing 3D space in working memory: Spatial images from vision, hearing, touch, and language. In: S. Lacey and R. Lawson (eds.) *Multisensory imagery* (pp. 131-155). Springer New York. doi: 10.1007/978-1-4614-5879-1_8

Lopez, C. (1997). *Going Faster! Mastering the Art of Race Driving: The Skip Barber Racing School*. Bentley Publishers.

Macadam, C. C. (2003). Understanding and Modeling the Human Driver. *Vehicle System Dynamics*, 40, 101–134. doi: 10.1076/vesd.40.1.101.15875

McGuire, L. M., & Sabes, P. N. (2009). Sensory Transformations and the Use of Multiple Reference Frames for Reach Planning. *Nature Neuroscience*, 12(8), 1056-1061.

McRuer, D. T., Allen, R. W., Weir, D. H., & Klein, R. H. (1977). New Results in Driver Steering Control Models. *Human Factors*, 19, 381–397. doi: 10.1177/001872087701900406

Menon, V., & Uddin, L. Q. (2010). Saliency, Switching, Attention and Control: a Network Model of Insula Function. *Brain Structure and Function*, 214(5-6), 655-667.

Miall, R. C., & Wolpert, D. M. (1996). Forward Models for Physiological Motor Control. *Neural networks*, *9*(8), 1265-1279.

Michon, J. A. (1985). A Critical View of Driver Behaviour Models: What Do We Know, What Should We Do?. In *Human behaviour and traffic safety* (pp. 485-524). Springer, Boston, MA.

Milliken, W.F. & Milliken, D.F. (1995). *Race Car Vehicle Dynamics*. Society of Automotive Engineers.

Moran, A., Collet, C. , Guillot, A., Campbell, M., Matthews J., Mahoney, C. & Lowther, J. (2013).The BASES Expert Statement on the Use of Mental Imagery in Sport, Exercise and Rehabilitation Contexts. *The Sport and Exercise Scientist*, 38: 10-11.

Moss, S. & Nye, D. (1987). *Stirling Moss, My Cars My Career*. Patrick Stephens.

Mnih, V., Kavukcuoglu, K., Silver, D., Rusu, A.A., Veness, J., Bellemare, M.G., ... & Petersen, S. (2015). Human-level Control Through Deep Reinforcement Learning. *Nature*, 518(7540), 529.

Mumford, D. (1992). On the Computational Architecture of the Neocortex. *Biological cybernetics*, *66*(3), 241-251.

Mundale, J. (2001). The Neuroanatomical Foundations of Cognition: Connecting the Neuronal Level wih the Study of Higher Brain Areas. In: W. Bechtler, P. Mandik, J. Mundale & R.S. Stufflebeam (eds.): *Philosophy and the Neurosciences, A Reader*. Blackwell.

Münte, T. F., Altenmüller, E., & Jäncke, L. (2002). The Musician's Brain as a Model of Neuroplasticity. *Nature Reviews Neuroscience*, *3*(6), 473-478.

Navarro, J., Reynaud, E., & Osiurak, F. (2018). Neuroergonomics of Car Driving: A Critical Meta-analysis of Neuroimaging Data on the Human Brain behind the Wheel. *Neuroscience & Biobehavioural Reviews*, *95*, 464-479.

Nash, C. J., Cole, D. J., & Bigler, R. S. (2016). A Review of Human Sensory Dynamics for Application to Models of Driver Steering and Speed Control. *Biological Cybernetics*, 110(2–3), 91–116. doi: 10.1007/ s00422-016-0682-x

Nastase, S. A., Gazzola, V., Hasson, U., & Keysers, C. (2019). Measuring Shared Responses across Subjects using Intersubject Correlation. *Social Cognitive & Affective Neuroscience*, *14*(6), 669-687.

Näätänen, R. and H. Summala (1976). *Road-user Behaviour and Traffic Accidents*, North-Holland Publishing Company, Amsterdam.

Newell, A. (1973). You Can't Play 20 Questions with Nature and win: Projective Comments on the Papers of this Symposium. In W. G. Chase (Ed.), *Visual information processing*.New York: Academic Press. pp. 203-308.

Newell, A., & Rosenbloom, P. S. (1982). *Mechanisms of Skill Acquisition and the Law of Practice*. Pittsburgh, PA: Design Research Center, Carnegie-Mellon University.

Niv, Y. (2009). Reinforcement Learning in the Brain. *Journal of Mathematical Psychology* 53: 139-154. doi:10.1016/j.jmp.2008.12.005

Niv, Y. & Schoenbaum, G. (2008). Dialogues on Prediction Errors. *Trends in Cognitive Sciences*, 12(7), 265-272. doi:10.1016/j.tics.2008.03.00

Nummenmaa, L., Lahnakoski, J. M., & Glerean, E. (2018). Sharing the Social world via Intersubject Neural Synchronisation. *Current Opinion in Psychology*, *24*, 7-14.

Palmeri, J. T. (1999). Theories of Automaticity and the Power Law of Practice. *Journal of Experimental Psychology*, 25, 543–551. doi: 10.1037/0278-7393.25.2.543

Petrides, M. (2005). Lateral Prefrontal Cortex: Architectonic and Functional Organization. *Philosophical Transactions of the Royal Society B: Biological Sciences*, 360 (1456), 781-795. doi:10.1098/rstb.2005.1631

Petzschner, F.H., Glasauer, S. & Stephan, K.E. (2015). A Bayesian Perspective on Magnitude Estimation. *Trends in Cognitive Sciences*, 19, 5, 285-293. doi: 10.1016/j.tics.2015.03.002

Posner, M. I., & Cohen, Y. (1984). Components of Visual Orienting. *Attention and Performance X: Control of Language Processes*, *32*, 531-556.

Postle, R.W. (2015): *Essentials of Cognitive Neuroscience*. Wiley-Blackwell.

Potkanowicz, E.S. & Mendel, R.W. (2013). The Case for Driver Science in Motorsport: A Review and Recommendations. *Sports Medicine*. DOI 10.1007/s40279-013-0040-2

Potkanowicz, E.S. (2019). The Physiological Stress of Automobile Racing. In: D.P. Ferguson (ed.) *The Science of Motorsport*. Routledge.

Pouget, A., & Sejnowski, T. (1997). Spatial Transformations in the Parietal Cortex Using Basis Functions. *Journal of Cognitive Neuroscience*, 9(2), 222-237.

Profeta, V. L., & Turvey, M. T. (2018). Bernstein's Levels of Movement Construction: A Contemporary Perspective. *Human movement science*, *57*, 111-133.

Prost, A. & Rousselot, P.-F. (1990). *Competition Driving*. Hazleton Publishing.

Ptak, R. (2012). The Frontoparietal Attention Network of the Human Brain: Action, Saliency, and a Priority Map of the Environment. *The Neuroscientist*, *18*(5), 502-515.

Raghavan, R.T, Prevosto, V. & Sommer, M.A. (2016). Contribution of Cerebellar Loops to Action Timing. *Current Opinion in Behavioural Sciences*, 8, 28–34. doi: 10.1016/j.cobeha.2016.01.008

Raichle, M. E., MacLeod, A. M., Snyder, A. Z., Powers, W. J., Gusnard, D. A., & Shulman, G. L. (2001). A Default Mode of Brain Function. *Proceedings of the National Academy of Sciences*, *98*(2), 676-682.

Raichle, M. E. (2015). The Brain's Default Mode Network. *Annual review of neuroscience*, *38*, 433-447.

Ramnani, N. (2006). The Primate Cortico-cerebellar System: Anatomy and Function. *Nature Reviews Neuroscience*, 7(7):511 - 522. doi: 10.1038/nrn1953

Rao, R. P., & Ballard, D. H. (1999). Predictive Coding in the Visual Cortex: a Functional Interpretation of Some Extra-classical Receptive-field Effects. *Nature Neuroscience*, *2*(1), 79-87.

Raudies, F., Mingolla, E., & Neumann, H. (2012). Active Gaze Control Improves Optic Flow Based Segmentation and Steering. *PloS ONE*, 7(6), e38446.

Raudies, F., Mingolla, E., & Neumann, H. (2012). Active Gaze Control improves Optic Flow Based Segmentation and Steering. *PloS ONE*, 7(6), e38446.

Reid, M. B., & Lightfoot, J. T. (2019). The Physiology of Auto Racing. *Medicine and science in sports and exercise*, *51*(12), 2548-2562.

Remonda, A., Veas, E., Luzhnica, G. (2021). Comparing Driving Behaviour of Humans and Autonomous Driving in a Professional Racing Simulator. PLoS ONE 16(2): e0245320 doi: 10.1371/journal.pone.0245320

Rizzolatti, G., Riggio, L., Dascola, I., & Umiltá, C. (1987). Reorienting Attention across the Horizontal and Vertical Meridians: Evidence in Favor of a Premotor Theory of Attention. *Neuropsychologia*, *25*(1), 31–40. doi: 10.1016/0028-3932(87)90041-8

Rizzolatti, G., & Craighero, L. (1998). Spatial Attention: Mechanisms and Theories. *Advances in Psychological Science*, *2*, 171-198.

Rosch, E. H. (1973). Natural Categories. *Cognitive Psychology*, 4 (3): 328–350.

Schmahmann, J.D. (1991). An Emerging Concept, the Cerebellar Contribution to Higher Function. *Archives of Neurololgy*, 48, 1178–1187.

Schmidt, R. A. (1975). A Schema Theory of Discrete Motor Skill Learning. *Psychological review*, *82*(4), 225.

Schmidt, R. A. (2003). Motor Schema Theory after 27 Years: Reflections and Implications for a New Theory. *Research quarterly for exercise and sport*, *74*(4), 366-375.

Schuster, C., Hilfiker, R., Amft, O., Scheidhauer, A., Andrews, B., Butler, J., Kischka, U. & Ettlin, T. (2011). Best Practice for Motor Imagery: a Systematic Literature Review on Motor Imagery Training Elements in Five Different Disciplines. *BMC Medicine*, 9, 75. doi:10.1186/1741-7015-9-75

Scolari, M., Seidl-Rathkopf, K. N., & Kastner, S. (2015). Functions of the Human Frontoparietal Attention Network: Evidence from Neuroimaging. *Current opinion in behavioural sciences*, *1*, 32-39.

Segers, J. (2014). *Analysis Techniques for Racecar Data Acquisition*. 2nd, edition. SAE International.. Warrendale, PA.

Senders, J. W., Kristofferson, A. B., Levison, W. H., Dietrich, C. W., & Ward, J. L. (1967). The Attentional Demand of Automobile Driving. *Highway Research Record*, (195).

Senna, A. (1993). Ayrton Senna's Principles of Race Driving. Hazleton Publising.

Sharp, R. S., & Peng, H. (2011). Vehicle Dynamics Applications of Optimal Control Theory. *Vehicle System Dynamics*, 49, 1073–1111. doi: 10.1080/00423114.2011.586707

Shea, C. H., & Wulf, G. (2005). Schema Theory: A Critical Appraisal and Re-evaluation. *Journal of Motor Behaviour, 37*(2), 85-102.

Shi, Z. & Burr, D. (2016). Predictive Coding of Multisensory Timing. *Current Opinion in Behavioural Sciences*, , 8, 200–206. doi: 10.1016/j.cobeha.2016.02.014

Silver, D., Hubert, T., Schrittwieser, J., Antonoglou, I., Lai, M., Guez, A., ... & Lillicrap, T. (2018). A general reinforcement learning algorithm that masters chess, shogi, and Go through self-play. *Science*, 362(6419), 1140-1144.

Simon, H.A. (1955). A Behavioural Model of Rational Choice. *Quarterly Journal of Economics*, 69, 99–118.

Simon, H. A. (1974). How Big is a Chunk?: By Combining Data from Several Experiments, a Basic Human Memory Unit Can be Identified and measured. *Science, 183*(4124), 482-488.

Smith, C. (1996). *Drive to Win, the Essential Guide to Race Driving*. Carroll Smith Consulting inc.

Smith, K. & Hancock, P.A. (1995). Situation Awareness is Adaptive, Externally Directed Consciousness. *Human Factors*, 37(1), 137-148.

Spick, M. (1988). The *Ace Factor. Air Combat and the Role of Situational Awareness.* Naval Institute Press.

Summala, H. (2007). Towards understanding motivational and emotional factors in driver behaviour: comfort through satisficing. In: P.Cacciabue (ed.) *Modelling Driver Behaviour in Automotive Environments*, pp. 189-207, Springer Verlag, London.

Summers, J. J., & Anson, J. G. (2009). Current status of the motor program: Revisited. *Human Movement Science, 28*(5), 566-577.

Taruffi, P. (1959). *The Technique of Motor Racing*. Bentley Publishers.

Tatler, B. W., & Land, M. F. (2011). Vision and the representation of the surroundings in spatial memory. *Philosophical Transactions of the Royal Society B: Biological Sciences,* 366(1564), 596-610. doi:10.1098/rstb.2010.0188

Ting, L. H., & McKay, J. L. (2007). Neuromechanics of muscle synergies for posture and movement. *Current opinion in neurobiology, 17*(6), 622-628.

Treisman, A. (1996). The Binding Problem. *Current Opinion in Neuro-biology, 6*(2), 171-178.

van der Linden D., Tops, M. & Bakker, A.B. (2020). Go with the Flow – A Neuroscientific View on Being Fully Engaged. *European Journal of Neuroscience*, 00, pp. 1-17. DOI: 10.1111/ejn.15014

van Leeuwen, P.M., de Groot, S., Happee, R. & de Winter. J.C.F. (2017). Differences between Racing and Non-racing Drivers: A Simulator Study Using Eye-tracking. *PLoS ONE,* 12(11): e0186871. doi: 10.4121/uuid:9ab22c4c-1c68-4838-a752-e4ba4ddeec07

Vansteenkiste, P., Cardon, G., D'Hondt, E., Philippaerts, R., & Lenoir, M. (2013). The Visual Control of Bicycle Steering: The Effects of Speed and Path Width. *Accident Analysis & Prevention*, 51, 222-227.

Watkins, E.S. (2006). The Physiology and Pathology of Formula One Grand Prix Motor Racing. *Clinical Neurosurgery*, 53, 145-152.

Wolfe, B., Dobres, J., Rosenholtz, R. & Reimer, B. (2017). More than the Useful Field: Considering Peripheral Vision in Driving. *Applied Ergonomics*, 65, 316-325. doi: 10.1016/j.apergo.2017.07.009

Wolpert, D. M., & Kawato, M. (1998). Multiple paired forward and inverse models for motor control. *Neural networks, 11*(7), 1317-1329.

Yeo, B. T., Krienen, F. M., Sepulcre, J., Sabuncu, M. R., Lashkari, D., Hollinshead, M., ... & Buckner, R. L. (2011). The organization of the human cerebral cortex estimated by intrinsic functional connectivity. *Journal of Neurophysiology*, 1125-1165.

IMAGE CREDITS

Stock photo credits (licenced through Shutterstock):

Figure 1-1	Cristiano Barni
Figure 2-1	Cristiano Barni
Figure 3-1	Cristiano Barni
Figure 4-1	Ivan Garcia
Figure 5-3	Nilovsergey
Figure 5-5	Veronica Louro
Figure 6-1	Veronica Louro
Figure 9-1	Ev. Safronov
Figure EN-1	Cristiano barni, Rodrigo Garrido & Francesc Juan

Brain images by SciePro / Shutterstock used in:

3-3	4-4	6-1	6-2	6-3	6-4	6-5	6-6	6-7
7-1	8-1	8-2	9-2	front cover				

Cover portrait of OL by Markku Verkasalo

Front cover typeface: Big Noodle Titling by James Arboghast

INDEX

E

F

G

H

I

J

K

L

T

U

V

W

Y

Z

ABOUT THE AUTHORS

Otto Lappi PhD is an Adjunct Professor of Cognitive Science and a Research Fellow of the Academy of Finland. He leads a research unit at the University of Helsinki with a 50-year track record of experimental work in naturalistic & laboratory conditions as well as driving simulators. He lives in the south of Finland.

Alan Dove is a Race Driver Performance analyst from with experience of coaching drivers across multiple racing disciplines. Having raced karts since the age of 8, he has had a lifetime of observing the development of the best drivers in the world.

Find FREE bonus content

and contact us at:

www.racersbrain.org

www.ingramcontent.com/pod-product-compliance
Ingram Content Group UK Ltd.
Pitfield, Milton Keynes, MK11 3LW, UK
UKHW021829190726
13853UKWH00003B/1269

9 789529 45854